Conjecture and Proof

© 2001 by
The Mathematical Association of America (Incorporated)
Library of Congress Catalog Card Number 2001095773

ISBN 0-88385-722-7

Printed in the United States of America

Current Printing (last digit):
10   9   8   7   6   5   4   3   2   1

Conjecture and Proof

Miklós Laczkovich

Published by
THE MATHEMATICAL ASSOCIATION OF AMERICA

CLASSROOM RESOURCE MATERIALS

This series provides supplementary material for students and their teachers—laboratory exercises, projects, historical information, textbooks with unusual approaches for presenting mathematical ideas, career information, and much more.

101 Careers in Mathematics, edited by Andrew Sterrett
Archimedes: What Did He Do Besides Cry Eureka?, Sherman Stein
Calculus Mysteries and Thrillers, R. Grant Woods
Combinatorics: A Problem Oriented Approach, Daniel A. Marcus
A Course in Mathematical Modeling, Douglas Mooney and Randall Swift
Cryptological Mathematics, Robert Edward Lewand
Elementary Mathematical Models, Dan Kalman
Geometry From Africa: Mathematical and Educational Explorations, Paulus Gerdes
Interdisciplinary Lively Application Projects, edited by Chris Arney
Laboratory Experiences in Group Theory, Ellen Maycock Parker
Learn from the Masters, Frank Swetz, John Fauvel, Otto Bekken, Bengt Johansson, and Victor Katz
Mathematical Modeling in the Environment, Charles Hadlock
A Primer of Abstract Mathematics, Robert B. Ash
Proofs Without Words, Roger B. Nelsen
Proofs Without Words II, Roger B. Nelsen
A Radical Approach to Real Analysis, David M. Bressoud
She Does Math!, edited by Marla Parker
Solve This: Math Activities for Students and Clubs, James S. Tanton

MAA Service Center
P.O. Box 91112
Washington, DC 20090-1112
1-800-331-1MAA   FAX: 1-301-206-9789

Introduction

This text is an elaborate version of the lecture notes for a one-semester course of the Budapest Semesters in Mathematics (BSM), a program for American and Canadian students. This program was initiated and designed by Paul Erdős, László Lovász, Vera T. Sós and László Babai in 1983–84 with the intention to offer undergraduate courses conveying the tradition of Hungarian mathematics. It was László Babai who proposed, in the spirit of this tradition, a course based on creative problem solving. He also invented the title "Conjecture and Proof", referring to Paul Erdős' slogan, or imperative "conjecture and prove!".

When, upon László Babai's request, I collected the material for the course, I realized that, even if it places emphasis on creative problem solving, the course must also contain some lectures discussing the results and methods of certain areas of mathematics. I selected the topics and collected the problems in 1984, and gave the course during the first semester of BSM in 1985. It proved to be successful, and thus "Conjecture and Proof" became a standard course of BSM. I also gave the course in 1986 at St. Olaf College. Since 1986, on most occasions, the instructor of the course has been Prof. G. Elekes.

In the selection of topics my intention was to give miniature introductions to various areas of mathematics, and present some interesting and important, but easily accessible results and methods. However, "easily accessible" does not mean "elementary": the text contains deep theorems with complete proofs, like the transcendence of e, the Banach–Tarski paradox or the existence of Borel sets of arbitrary (finite) class. One of my purposes was to demonstrate how far we can get from first principles in just a couple of steps.

Although the text discusses questions from various fields including number theory, algebra and geometry, it is centered around the real number system and the problem of measure. Thus the number theoretic sections are concerned with rational and irrational and with algebraic and transcendental numbers; the problems of geometric constructions clarify the nature of constructible numbers (as a subset of algebraic numbers), and the questions of geometric dissections serve as motivation for general problems of equidecomposability. This also illustrates the great diversity of mathematics on the one hand, and the (sometimes surprising) connections between apparently independent areas of mathematics on the other hand.

In short, this text attempts to give "an introduction to the spirit of mathematics"*. As for the necessary prerequisites, I tried to keep them at the minimum. We shall use, without explanation, the binomial theorem, complex numbers, mathematical induction, the notion of continuity, and integration. (In fact, the use of integration is restricted to three instances in Sections 1 and 5.) That is, any introductory calculus course provides much of the necessary background for understanding this book. The logical dependence of the sections is the following. The first two sections are self-contained. Section 3 uses a particular result from Section 4. Other notions and results from Section 4 are needed in Sections 5 and 6, while Section 7 is built on Section 6. Some parts of Sections 1, 8, and 9 are closely related. Sections 10 and 11 are self-contained and serve as a basis for the next two sections. Section 16 uses the previous two sections, and is needed for the last section.

This book differs from the "Conjecture and Proof" course in one respect: it uses different exercises (the original problems had to be reserved for the course). Still, I hope that the exercises given at the end of each section will prove instructive.

By introducing a variety of advanced topics, the book functions, in part, as a survey of topics from number theory, geometry, measure theory, and set theory. It can be used as a supplement in courses that introduce abstract mathematics to undergraduates. The ideas that are presented are deeper and more sophisticated than those typically encountered in sophomore-level "transition" courses. However, talented students in such courses should find this book to be an exciting excursion into new areas of mathematics—and more importantly, new ways of thinking about mathematical problems. Because of its unusual depth and the fact that some of the sections can stand alone or

---

* This is the subtitle of S. Stein's book *Mathematics, the Man-Made Universe* [15].

be combined with a few others to form a unit, this book is ideally suited for upper-level undergraduate seminars or capstone courses.

Acknowledgment. I would like to thank Prof. G. Elekes who read the manuscript meticulously, and offered several suggestions and corrections. His help in the writing of this text was invaluable.

Miklós Laczkovich
June 14, 2001

Contents

Part I

Proofs of Impossibility, Proofs of Nonexistence

The reason we start with proofs of impossibility is that they provide the best introduction to the "spirit of mathematics". Indeed, when we prove that something is impossible, a certain problem is unsolvable, or certain objects and procedures do not exist, our arguments are always very abstract, unambiguous and final. Mark Kac and Stanislaw Ulam write ([7], p. 26):

> "The unique and peculiar character of mathematical reasoning is best exhibited in proofs of impossibility. When it is asserted that doubling the cube (i.e., constructing $\sqrt[3]{2}$ with a ruler and a compass) is impossible, the statement does not merely refer to a *temporary* limitation of human ability to perform this feat. It goes far beyond this, for it proclaims that *never*, no matter what, will anybody ever be able to construct $\sqrt[3]{2}$ or to trisect a general angle if the only instruments at his disposal are a straightedge and a compass.
> No other science, or for that matter no other discipline of human endeavour, can even contemplate anything of such finality."

We start with proofs of irrationality, since they are the prototypes of proofs of impossibility: in these arguments we have to prove that it is *impossible* to represent a given number as the quotient of integers.

The question whether a given number is rational or not is among the simplest, most natural, and most difficult problems. It is still not known if the numbers $e + \pi$, $e \cdot \pi$ or Euler's constant

$$\gamma = \lim_{n \to \infty} \left(1 + \frac{1}{2} + \cdots + \frac{1}{n} - \log n \right)$$

1

are rational or not. This explains the necessity of advanced tools when dealing with some of these problems. In the proof of the irrationality of π or the transcendence of e, we shall use integration. These will be the only instances when we use integrals.

The investigations concerning questions of irrationality were initiated by ancient Greek mathematics, as were the questions concerning geometric constructions. We discuss the latter problems in Sections 2 and 3. The necessary methods (elements of linear algebra and field extensions) will be explained in Section 4. In Section 7 we shall consider the impossibility of some geometric dissections. The tools needed for these proofs of impossibility (Hamel bases and Cauchy's functional equation) are discussed in Section 6.

1

Proofs of Irrationality

A real number is said to be rational if it can be expressed as the quotient of two integers; otherwise it is called irrational. One of the basic discoveries of ancient Greek mathematics is that irrational quantities indeed exist. For example, it was already known by the Pythagorean school that $\sqrt{2}$ *is irrational.* Due to the importance of this fact, we shall give five proofs.

1. Suppose $\sqrt{2} = p/q$, where p, q are positive integers and let q be the smallest such number. Then $2q^2 = p^2$ and thus p^2 is even. Then p itself must be even; let $p = 2p_1$. Then $2q^2 = (2p_1)^2 = 4p_1^2$, $q^2 = 2p_1^2$ and thus q is also even. If $q = 2q_1$ then $\sqrt{2} = p/q = p_1/q_1$. Since $q_1 < q$, this contradicts the minimality of q.

2. Suppose again that $\sqrt{2} = p/q$ where p, q are positive integers and q is the smallest possible. Then we have

$$\frac{2q - p}{p - q} = \frac{2 - (p/q)}{(p/q) - 1} = \frac{2 - \sqrt{2}}{\sqrt{2} - 1} = \sqrt{2}.$$

Since $2q - p$ and $p - q$ are integers and $0 < p - q < q$, this again contradicts the minimality of q.

3. This is a geometric proof (see Figure 1). Let $ABC_\triangle$ be an isosceles right triangle; then, by Pythagoras' theorem, the ratio of the segments BC and AB is $\sqrt{2}$. If $\sqrt{2} = p/q$ where p, q are positive integers, then the segments BC and AB are commensurable; that is, they are both integral

3

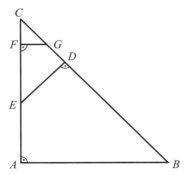

FIGURE 1

multiples of the same length. Indeed, if $d = \overline{AB}/q$ then $\overline{AB} = q \cdot d$ and $\overline{BC} = \sqrt{2} \cdot \overline{AB} = (p/q) \cdot qd = p \cdot d$.

Now let D be the point of the hypotenuse BC such that $\overline{BD} = \overline{AB}$, and let the perpendicular to BC at D intersect AC at the point E. It is easy to see that $\overline{AE} = \overline{ED} = \overline{DC}$. Thus $\overline{CD} = \overline{BC} - \overline{AB}$ and $\overline{EC} = \overline{AC} - \overline{AE} = \overline{AB} - (\overline{BC} - \overline{AB})$ are both integer multiples of d (since $\overline{AB}$ and $\overline{BC}$ are).

We repeat this argument for the triangle $EDC_\triangle$: if $\overline{EF} = \overline{ED}$ and if the perpendicular to EC at F meets DC in G, then $\overline{FG}$ and $\overline{GC}$ are both integer multiples of d, since $\overline{ED}$ and $\overline{EC}$ are. Repeating this argument indefinitely, we obtain a strictly decreasing sequence of segments (AC, EC, FC, etc.) such that each of them is an integral multiple of the same length d. Then the corresponding positive integers constitute a strictly decreasing sequence, which is impossible.

4. This proof uses the *fundamental theorem of arithmetic*. This theorem states that every integer $n > 1$ can be expressed as a product of primes, and this representation is unique, apart from the order of the factors. If $\sqrt{2} = p/q$ then $2q^2 = p^2$. In the representation of p^2 as a product of primes, every prime will occur an even number of times. On the other hand, in the representation of $2q^2$ the prime 2 will occur an odd number of times. Since the representation is unique, this is impossible.

5. Let $\varepsilon_i = (\sqrt{2}-1)^i$. Then $0 < \varepsilon_i < 1/2^i$ for every i, since $0 < \sqrt{2}-1 < 1/2$. It is easy to see, using induction on i, that $\varepsilon_i = a_i + b_i\sqrt{2}$, where a_i and b_i are integers. (This also follows at once from the binomial theorem.) If

$\sqrt{2} = p/q$ then

$$\varepsilon_i = a_i + b_i \cdot \frac{p}{q} = \frac{a_i q + b_i p}{q} = \frac{A_i}{q},$$

where A_i is an integer. Since $\varepsilon_i > 0$, this gives $A_i \geq 1$ and $\varepsilon_i \geq 1/q$. Therefore $1/q \leq \varepsilon_i \leq 1/2^i$ for every i, which is impossible, since $2^i > 2^q > q$ if $i > q$.

The first three of these proofs only use elementary principles. The next two are somewhat more advanced. The second and third proofs are connected: if $\overline{BC} = p$ and $\overline{AB} = q$ then $\overline{EC} = 2\overline{AB} - \overline{BC} = 2q - p$ and $\overline{ED} = \overline{BC} - \overline{AB} = p - q$. By Pythagoras' theorem we have $(2q-p)/(p-q) = p/q = \sqrt{2}$; we applied the same relation in the second proof.

The irrationality of $\sqrt{2}$ is the special case of the following theorem: *if n is any positive integer, then $\sqrt{n}$ is either an integer or irrational.* This can be proved by modifying some of the proofs above. The first proof works only if n is even but is not a multiple of 4. A modification of the second proof can be applied for every n. Indeed, let $k = [\sqrt{n}]$. If $\sqrt{n}$ is not an integer, then $\sqrt{n} = p/q$ implies $(nq - kp)/(p - kq) = \sqrt{n}$, and $0 < p - kq < q$. A simple version of the third proof works for every n of the form $m^2 + 1$. The fourth proof works for every n. The same is true for the fifth proof. Indeed, if $k = [\sqrt{n}]$, then $\varepsilon_i = (\sqrt{n} - k)^i$ is of the form $a_i + b_i \sqrt{n}$, where a_i, b_i are integers. If $\sqrt{n}$ is not an integer, then $0 < \varepsilon_i \to 0$ as $i \to \infty$, and this gives a contradiction in the same way as in the case of $\sqrt{2}$.

We can generalize further: *if n, k are positive integers, then $\sqrt[k]{n}$ is either an integer or irrational.* The fourth and the fifth proofs work. On the other hand, it is not known whether or not an argument similar to the second proof can be applied. This is unsolved even for $\sqrt[3]{2}$.

The number e is defined as the limit of the sequence $\left(1 + \frac{1}{n}\right)^n$. This constant occurs frequently in calculus. One can show that $e = 1 + \frac{1}{1!} + \frac{1}{2!} + \cdots$, giving it the value $2.718\ldots$. Another important property of e is that the derivative of the exponential function e^x is itself: $(e^x)' = e^x$.

The number e is irrational.

We give two proofs.

1. Suppose that $e = p/q$, where p, q are positive integers. Then $q > 1$ and

$$q!\left(e - 1 - \frac{1}{1!} - \frac{1}{2!} - \cdots - \frac{1}{q!}\right) = q!\left(\frac{1}{(q+1)!} + \frac{1}{(q+2)!} + \cdots\right). \quad (1)$$

The left-hand side is an integer. We shall arrive at a contradiction by showing that the right-hand side of (1) is not an integer, since it is positive, but is strictly less than 1. Indeed, the right-hand side of (1) equals

$$\frac{1}{(q+1)} + \frac{1}{(q+1)(q+2)} + \cdots < \frac{1}{(q+1)} + \frac{1}{(q+1)^2} + \cdots = \frac{1}{q} < 1.$$

2. We shall use an argument somewhat similar to the fifth proof of the irrationality of $\sqrt{2}$. Let $I_n = \int_0^1 x^n e^x \, dx$. First we show that for every $n = 0, 1, \ldots$ there are integers a_n and b_n such that $I_n = a_n + b_n e$.

We shall prove our statement by induction. We have $I_0 = \int_0^1 x^0 e^x \, dx = \int_0^1 e^x \, dx = [e^x]_0^1 = e - 1$, and thus the statement is true for $n = 0$ (take $a_0 = -1$, $b_0 = 1$). If $n \geq 1$ and $I_{n-1} = a_{n-1} + b_{n-1} e$, then integration by parts gives

$$I_n = \int_0^1 (e^x)' x^n \, dx = [e^x x^n]_0^1 - \int_0^1 e^x \cdot n x^{n-1} dx = e - n \cdot I_{n-1}$$

$$= e - n \cdot (a_{n-1} + b_{n-1} e) = -n a_{n-1} + (1 - n b_{n-1}) e,$$

which proves our statement concerning I_n.

Now suppose that $e = p/q$, where p, q are positive integers. Then

$$0 < I_n = a_n + b_n \frac{p}{q} = \frac{a_n q + b_n p}{q}.$$

This implies that $a_n q + b_n p \geq 1$ and thus $I_n \geq 1/q$. On the other hand, $e^x \leq e$ for every $x \in [0, 1]$ and hence $I_n \leq \int_0^1 x^n \cdot e \, dx = e/(n+1)$. Therefore $1/q \leq e/(n+1)$ for every n, which is clearly impossible.

Our next aim is to show that π *is irrational*. Actually, we shall prove the following more general statement. *If $0 < r \leq \pi$ is rational, then either* $\sin r$ *or* $\cos r$ *is irrational*. This easily implies the irrationality of π. Indeed, if π were rational, then one of $\sin \pi = 0$ and $\cos \pi = -1$ would be irrational which is not the case.

We shall use an argument analogous to the second proof of the irrationality of e. In this case we shall investigate the integral $J_n = \int_0^1 f(x) \sin rx \, dx$, where $f(x) = \frac{1}{n!} x^n (1 - x)^n$.

First we prove that *if g is a polynomial with integer coefficients and $h(x) = x^n g(x)/n!$, then $h^{(k)}(0)$ is an integer for every k.*

Indeed, $h(x)$ is of the form $\sum_{i=n}^m (c_i/n!) x^i$, where $c_n, \ldots, c_m$ are integers. Hence, $h^{(k)}(0) = 0$ for every $k < n$ and $k > m$. On the other hand, if $n \leq k \leq m$, then $h^{(k)}(0) = (c_k/n!) \cdot k! = c_k (n+1) \cdots k$ is an integer, as we stated.

Applying this lemma to $g(x) = (1-x)^n$ we obtain that the numbers $f^{(k)}(0)$ are integers for every k. Since $f(1-x) = f(x)$, we have $f^{(k)}(1) = (-1)^k f^{(k)}(0)$, and thus the numbers $f^{(k)}(1)$ are also integers.

Using integration by parts successively $2n$ times we obtain

$$J_n = -\left[f(x)\frac{\cos rx}{r}\right]_0^1 + \frac{1}{r}\int_0^1 f'(x)\cos rx\, dx$$

$$= -\frac{1}{r}[f(1)\cos r - f(0)] + \frac{1}{r}\left[f'(x)\frac{\sin rx}{r}\right]_0^1 - \frac{1}{r^2}\int_0^1 f''(x)\sin rx\, dx$$

$$= -\frac{1}{r}[f(1)\cos r - f(0)] + \frac{1}{r^2}f'(1)\sin r + \frac{1}{r^3}[f''(1)\cos r - f''(0)]$$

$$- \cdots \pm \frac{1}{r^{2n+1}}\left[f^{(2n)}(1)\cos r - f^{(2n)}(0)\right]. \tag{2}$$

Suppose that $0 < r \le \pi$ and that the numbers r, $\sin r$, $\cos r$ are all rational. Let q be a common denominator of $1/r$, $\sin r$, and $\cos r$. Since $f^{(k)}(0)$ and $f^{(k)}(1)$ are integers, (2) implies that $J_n = A_n/q^{2n+2}$, where A_n is an integer. Now $f(x) > 0$ and $\sin rx > 0$ for every $x \in (0,1)$ (recall that $0 < r \le \pi$), which implies that $J_n > 0$. Therefore $A_n \ge 1$ and $J_n \ge 1/q^{2n+2}$.

On the other hand, $f(x)\sin rx \le 1/n!$ in $[0,1]$ and thus $J_n \le 1/n!$. Hence, $1/q^{2n+2} \le J_n \le 1/n!$; that is, $n! \le q^{2n+2} \le q^{3n}$ for every $n \ge 2$. However, for $n > 2q^6$ we have

$$n! > n(n-1)\cdots q^6 > \left(q^6\right)^{n/2} = q^{3n},$$

a contradiction.

We conclude this section by the investigation of the numbers $\cos r\pi$, where r is rational. Let $r = p/q$, where p and q are coprime integers and q is positive. We show that $\cos r\pi$ is rational if and only if $q = 1$, 2 or 3. The "if" part is clear since

$$\cos r\pi = \begin{cases} \pm 1 & \text{if } q = 1; \\ 0 & \text{if } q = 2; \text{ and} \\ \pm 1/2 & \text{if } q = 3. \end{cases}$$

In order to prove the "only if" part, first we show that for every $n = 0, 1, \ldots$ there is a polynomial $t_n(x)$ such that $2\cos nx = t_n(2\cos x)$ for every x. We shall argue by induction on n. For $n = 0$ and $n = 1$ we may take $t_0(x) \equiv 2$ and $t_1(x) \equiv x$. Let $n \ge 1$ and suppose that t_k exists for every $k \le n$. Then, applying the identity

$$\cos(n + 1)x + \cos(n - 1)x = 2 \cos nx \cos x$$

we obtain $2 \cos(n + 1)x = (2 \cos nx)(2 \cos x) - 2 \cos(n - 1)x$, and thus the polynomial $t_{n+1}(x) = t_n(x) \cdot x - t_{n-1}(x)$ satisfies $2 \cos(n + 1)x = t_{n+1}(2 \cos x)$. Therefore, t_n exists for every n.

Applying the recursion $t_{n+1} = t_n \cdot x - t_{n-1}$ it is easy to see by induction that, for every $n \geq 1$, t_n is of degree n, the leading coefficient of t_n is 1, and the other coefficients of t_n are also integers.

Now let $r \in \mathbf{Q}$ be such that $\cos r\pi$ is rational. Let $2 \cos r\pi = a/b$ where a and b are coprime integers. If the denominator of r is n, then nr is an integer and thus

$$\pm 2 = 2 \cos nr\pi = t_n(2 \cos r\pi) = t_n(a/b)$$
$$= (a/b)^n + c_{n-1}(a/b)^{n-1} + \cdots + c_0,$$

where $c_0, \ldots, c_{n-1}$ are integers. Multiplying by b^n we obtain

$$\pm 2b^n = a^n + c_{n-1}a^{n-1}b + \cdots + c_0 b^n.$$

Here every term is a multiple of b except a^n, therefore b divides a^n. Since $\gcd(a, b)$, the greates common divisor of a and b, is 1, this implies $b = \pm 1$ and thus $2 \cos r\pi$ is an integer. By $|2 \cos r\pi| \leq 2$, the possible values for $2 \cos r\pi$ are 0, ± 1 and ± 2. Thus $\cos r\pi = 0, \pm 1, \pm 1/2$ which implies either $r = k/2$ or $k/3$, where k is an integer. This completes the proof.

Exercises

1.1. Prove that the numbers

$$\cfrac{1}{1 + \cfrac{1}{1 + \cfrac{1}{\ddots + \cfrac{1}{1 + \cfrac{1}{1 + \sqrt{2}}}}}}$$

are irrational.

1.2. Prove that the number $\sqrt{2} + \sqrt{3}$ is irrational.

1.3. Define the sequences p_n and q_n by $p_1 = q_1 = 1$ and $p_{n+1} = 2q_n + p_n$, $q_{n+1} = p_n + q_n$ $(n = 1, 2, \ldots)$. Prove that $\lim_{n \to \infty} p_n/q_n = \sqrt{2}$.

1.4. Find a geometric proof of the irrationality of $\sqrt{5}$ and, more generally, that of $\sqrt{m^2 + 1}$ $(m = 1, 2, \ldots)$.

1.5. Prove that if $0 < r < \pi/2$ is rational, then $\tan r$ is irrational. (H) *

1.6. Prove that $\sin 1$ is irrational. (Here 1 is measured in radians. As for $\sin 1°$ see Exercise 1.8.) (H)

1.7. Prove that e^2 is irrational. (H)

1.8. Prove that if r and $\sin r\pi$ are both rational, then $\sin r\pi = 0, \pm 1, \pm 1/2$.

1.9. Prove that if r and $\tan r\pi$ are both rational, then $\tan r\pi = 0, \pm 1$.

2

The Elements of the Theory of Geometric Constructions

By a geometric construction we mean the application of the following procedure.

Given two points, A and B, in the plane, we construct one or two new points by

1) using a ruler to draw a straight line going through two of the points already constructed or given, or

2) using a compass to draw a circle such that its center is already constructed or given and its radius equals the distance of two points already constructed or given.

At each step we draw either two lines, or two circles, or one line and a circle in this way and construct their intersection(s). The construction consists of finitely many steps. Its aim is to construct two points with a given distance, two lines enclosing a given angle, or more generally, to construct the points of a given geometric figure. In the sequel we shall assume that the distance between the given points A and B is unity.

The following three classical problems were raised by Greek mathematicians more than 2500 years ago.

1. To construct a segment of length $\sqrt[3]{2}$ ("doubling the cube").

2. To give a general construction which trisects every given angle ("trisection of angles").

3. To construct a square having the same area as the unit circle ("squaring the circle").

The unsolvability of these problems was proven in the nineteenth century. In this section we shall consider the first two of these problems.

We fix a coordinate system such that A and B are the points $(0,0)$ and $(1,0)$, respectively. We shall say that a number d is *constructible*, if there is a construction that produces two points of distance $|d|$. It is clear that d is constructible if and only if there is a construction that produces the point $(d,0)$. In order to solve the problems above, we have to characterize the constructible numbers.

If the numbers a and b are constructible, then so are $a + b$ and $a - b$. Constructing suitable similar triangles, it is also easy to see that if a and $b \neq 0$ are constructible then so are ab and a/b. This immediately implies that every rational number is constructible.

Now we show that if a number $a > 0$ is constructible, then so is $\sqrt{a}$. Indeed, construct collinear points, P, Q and R such that $\overline{PQ} = a$, $\overline{QR} = 1$ and $\overline{PR} = a + 1$. Let C denote the circle with diameter PR, and let S be one of the intersections of C with the perpendicular to PR at the point Q. Then $\overline{QS} = \sqrt{a}$ (why?).

Therefore every number that can be obtained from rational numbers by applying finitely many additions, subtractions, multiplications, divisions, and taking square roots is constructible.

The converse is also true: if a number is constructible, then it can be obtained from the rationals applying the basic algebraic operations and taking square roots. Indeed, using the equations

$$(y - b_1)(a_2 - a_1) - (x - a_1)(b_2 - b_1) = 0$$

(the straight line going through the points (a_1, b_1) and (a_2, b_2)) and

$$(x - a)^2 + (y - b)^2 = r^2$$

(the circle with center (a, b) and radius r), an easy computation shows that the coordinates of the points produced by any construction can be obtained in this way.

We can express this characterization of constructible numbers in a more formal way by introducing the notion of fields.

Let F be a set of complex numbers. We say that F is a *field*, if $0, 1 \in F$ and $x, y \in F$ implies $x + y \in F$, $x - y \in F$, $xy \in F$, and, if $y \neq 0$, then $x/y \in F$. (For example, the set $\mathbf{Q}$ of rational numbers is a field. It is easy to see that every field contains $\mathbf{Q}$.)

If F is a field and $a \in F$ is a positive real number, then we shall denote by $F(\sqrt{a})$ the set of numbers of the form $x + y\sqrt{a}$, where $x, y \in F$. This is also a field. Indeed, it is clear that the sum, difference and product of two elements of $F(\sqrt{a})$ also belong to $F(\sqrt{a})$. To prove this for the quotient, it is enough to show that if $b \in F(\sqrt{a})$, $b \neq 0$, then $1/b \in F(\sqrt{a})$. This follows from the identity

$$\frac{1}{x + y\sqrt{a}} = \frac{x - y\sqrt{a}}{x^2 - ay^2},$$

taking into consideration that $x, y, a \in F$ implies

$$\frac{x}{x^2 - ay^2} \in F \quad \text{and} \quad \frac{-y}{x^2 - ay^2} \in F.$$

Let $F_0 = \mathbf{Q} \subset F_1 \subset \ldots \subset F_n$ be a sequence of fields, and suppose that there are positive numbers $a_0 \in F_0, \ldots, a_{n-1} \in F_{n-1}$ such that $F_i = F_{i-1}(\sqrt{a_{i-1}})$ for every $i = 1, \ldots, n$. Then the elements of F_n are constructible, since they can be obtained from the rationals using the basic operations and taking square roots. Conversely, if a number d is constructible, then, as we saw above, it can be obtained from the rationals using the basic operations and taking square roots, and this easily implies that there are fields F_i and positive numbers a_i with the properties described above and such that $d \in F_n$.

The real roots of a quadratic polynomial with rational coefficients are always constructible, since they belong to a field of the form $\mathbf{Q}(\sqrt{a})$ where $a \in \mathbf{Q}$. We show that the roots of cubic polynomials with rational coefficients are not always constructible. More precisely, we prove the following.

Let $f(x) = x^3 + px^2 + qx + r$ be a polynomial with $p, q, r \in \mathbf{Q}$, and suppose that f does not have a rational root. Then the real roots of f are not constructible.

Indeed, suppose that t is a constructible root of f. Then there is a sequence of fields $F_0 = \mathbf{Q} \subset F_1 \subset \cdots \subset F_n$ and there are positive numbers $a_i \in F_i$ such that $F_i = F_{i-1}(\sqrt{a_{i-1}})$ $(i = 1, \ldots, n)$, and $t \in F_n$. Then f has a root in F_n and does not have a root in $F_0 = \mathbf{Q}$. Therefore, there is a k such that f has a root in F_k and does not have a root in F_{k-1}. Then $F_k \neq F_{k-1}$ and there is an $a \in F_{k-1}$, $a > 0$ such that $F_k = F_{k-1}(\sqrt{a})$.

Let $y \in F_k$ be a root of f, and let $y = u + v\sqrt{a}$, where $u, v \in F_{k-1}$. We show that $\bar{y} = u - v\sqrt{a}$ is also a root of f. Indeed,

$$0 = f(y)$$
$$= (u + v\sqrt{a})^3 + p(u + v\sqrt{a})^2 + q(u + v\sqrt{a}) + r$$
$$= A + B\sqrt{a},$$

where

$$A = u^3 + 3uv^2a + pu^2 + pv^2a + qu + r \in F_{k-1}$$

and

$$B = 3u^2v + v^3a + 2puv + qv \in F_{k-1}$$

(note that $p, q, r \in \mathbf{Q}$). Since $\sqrt{a} \notin F_{k-1}$ by $F_k \neq F_{k-1}$, we have $A = B = 0$. Therefore $f(\overline{y}) = A - B\sqrt{a} = 0$, as we stated. Now y and $\overline{y}$ are different roots of f, since $y = \overline{y}$ would imply $y = u$, but $u \in F_{k-1}$ and hence u is *not* a root.

Let z be the third root of f. Then

$$f(x) = x^3 + px^2 + qx + r = (x - y)(x - \overline{y})(x - z),$$

and thus $y + \overline{y} + z = -p \in \mathbf{Q}$. Therefore $z = -p - (y + \overline{y}) = -p - 2u \in F_{k-1}$. Thus f has a root, z, in F_{k-1}, which is a contradiction.

The previous theorem solves the first classical problem: $\sqrt[3]{2}$ *is not constructible*. Indeed, $\sqrt[3]{2}$ is a root of $x^3 - 2$, and this cubic polynomial has rational coefficients but does not have any rational root.

We turn to the second problem. First note that an angle α can be constructed if and only if we can construct a right triangle with acute angle α and hypotenuse 1. Such a triangle, in turn, can be constructed if and only if the number $\cos\alpha$ is constructible. That is, the angle α is constructible if and only if $\cos\alpha$ is.

Now we show that $\cos 20°$ *is not constructible*. Indeed, since $2\cos 3\alpha = (2\cos\alpha)^3 - 3 \cdot (2\cos\alpha)$ and $2\cos 60° = 1$, we have $f(2\cos 20°) = 0$, where $f(x) = x^3 - 3x - 1$. This cubic polynomial has rational coefficients but does not have any rational root (why?), and thus $2\cos 20°$ is not constructible. Therefore the angle $20°$ cannot be constructed.

Thereby, the second classical problem is solved: *There is no general construction which trisects any given angle.* Indeed, otherwise there would be a construction for $20°$ (since there is one for $60°$).

As another immediate corollary we obtain that *the regular 9-gon cannot be constructed.* (Because $20°$ is the acute angle of the triangle determined by three consecutive vertices of a regular 9-gon.)

We conclude by showing that *the regular 7-gon cannot be constructed either.*

It suffices to show that $t = 2\cos\frac{2\pi}{7}$ is not constructible. Now $t = \varepsilon + \varepsilon^{-1}$, where $\varepsilon = \cos\frac{2\pi}{7} + i\sin\frac{2\pi}{7}$. Since

$$\frac{\varepsilon^7 - 1}{\varepsilon - 1} = \varepsilon^6 + \varepsilon^5 + \varepsilon^4 + \varepsilon^3 + \varepsilon^2 + \varepsilon + 1 = 0,$$

we have, dividing by ε^3,

$$(\varepsilon^3 + \varepsilon^{-3}) + (\varepsilon^2 + \varepsilon^{-2}) + (\varepsilon + \varepsilon^{-1}) + 1 = 0. \tag{3}$$

Now $t^3 = \varepsilon^3 + \varepsilon^{-3} + 3\varepsilon + 3\varepsilon^{-1}$ and $t^2 = \varepsilon^2 + \varepsilon^{-2} + 2$, and thus (3) gives $t^3 - 3t + t^2 - 2 + t + 1 = 0$. Therefore t is a root of the cubic polynomial $f(x) = x^3 + x^2 - 2x - 1 = 0$. Since f has rational coefficients but does not have rational roots, t is not constructible.

Exercises

2.1. Some of the following sets are fields. Identify them.

$$\{p + q\sqrt{2} + r\sqrt{3} : p, q, r \in \mathbf{Q}\},$$
$$\{p + q\sqrt{2} + r\sqrt{3} + s\sqrt{6} : p, q, r, s \in \mathbf{Q}\},$$
$$\{p + q\sqrt[3]{2} : p, q \in \mathbf{Q}\},$$
$$\{p + q\sqrt[3]{2} + r\sqrt[3]{4} : p, q, r \in \mathbf{Q}\}.$$

2.2. Prove that if a and b are positive rationals, then $\{p + q\sqrt{a} + r\sqrt{b} + s\sqrt{ab} : p, q, r, s \in \mathbf{Q}\}$ is a field. Generalize for arbitrary (even infinite) sets of positive rational numbers.

2.3. Prove that $\cos\alpha$ is constructible if and only if $\cos(\alpha/2)$ is constructible.

2.4. Prove that if $\cos\alpha$ and $\cos\beta$ are constructible, then so are $\cos(\alpha + \beta)$ and $\cos(\alpha - \beta)$.

2.5. Determine those integers k for which $\cos k° = \cos(k\pi/180)$ is constructible.

2.6. Prove that if $\alpha + \beta + \gamma = \pi$, then

$$1 - (\cos^2\alpha + \cos^2\beta + \cos^2\gamma) - 2\cos\alpha\cos\beta\cos\gamma = 0.$$

2.7. Let Δ be a triangle and let O denote the centre of the circumscribed circle of Δ. Suppose that the distances between O and the sides of Δ

are 1, 2 and 3. Prove that the triangle Δ cannot be constructed using ruler and compass. (H)

3

Constructible Regular Polygons

The exact characterization of the constructible regular polygons was given by C. F. Gauss in a celebrated theorem proved in 1801. Gauss' theorem is rather surprising in that it gives a purely number theoretic answer to a purely geometric problem.

The numbers $F_i = 2^{2^i} + 1$ $(i = 0, 1, \ldots)$ are called *Fermat numbers*. The first five Fermat numbers, corresponding to $i = 0, 1, 2, 3, 4$ are 3, 5, 17, 257 and 65537. Each of these numbers is prime. Based on this "evidence," Pierre Fermat conjectured (about 1640) that the numbers F_i are prime for every i. Fermat's conjecture was disproved by L. Euler in 1732; Euler discovered that 641 divides F_5 and thus F_5 is composite. The prime factorization of F_6 was found in 1880; it turned out that 274177 divides F_6. In 1970 it was shown that F_7 is the product of two primes consisting of 17 and 22 decimal digits, respectively. By now it is known that F_i is composite for every $5 \leq i \leq 23$. Several other Fermat numbers were also examined, but no Fermat primes have been found after F_4.

Gauss' theorem states that *the regular n-gon is constructible if and only if $n = 2^k p_1 \ldots p_m$, where $p_1, \ldots, p_m$ are different Fermat primes.*

Since 3 and 5 are Fermat primes, this implies that for $n = 3, 4, 5, 6, 8$ and 10 the regular n-gon is constructible (this was known already by Euclid). On the other hand, 7 is not a Fermat prime, and thus the regular 7-gon is not constructible (as we proved in the previous section). Also, the prime factorization of 9 contains a power of an odd prime, therefore the regular 9-gon is not constructible either.

In this section we sketch the proof of the "only if" part of Gauss' theorem. The proof is based on the notions of algebraic number and degree.

The set of polynomials with rational coefficients will be denoted by $\mathbf{Q}[x]$. A complex number α is said to be *algebraic* if it is the root of a nonzero polynomial $p \in \mathbf{Q}[x]$. Among the degrees of all nonzero polynomials $f \in \mathbf{Q}[x]$ satisfying $f(\alpha) = 0$, there is a minimal one. This minimal degree is called the *degree of the number* α.

For example, every rational number is algebraic of degree 1. The number $\sqrt{2}$ is algebraic of degree 2. The number $\sqrt[3]{2}$ is algebraic of degree 3 (prove it!).

The proof of Gauss' theorem is based on the following statement: *if a number is constructible, then it is algebraic, and its degree is a power of* 2.

We shall prove this in the next section; for the time being let us take it for granted. We show that if p is an odd prime and if the regular p-gon is constructible, then p must be a Fermat prime. Let $t = 2\cos\frac{2\pi}{p}$. Then t is constructible and thus t is algebraic and its degree is a power of 2. Since $t = \varepsilon + \varepsilon^{-1}$ where $\varepsilon = \cos\frac{2\pi}{p} + i\sin\frac{2\pi}{p}$, it follows that the degree of ε is also a power of 2 (the proof of this is also postponed to the next section; see Exercise 4.7). Now ε is a root of

$$\frac{x^p - 1}{x - 1} = x^{p-1} + x^{p-2} + \cdots + 1,$$

and hence the degree of ε is at most $p - 1$. It can be shown that the degree of ε is *exactly* $p - 1$ (we omit the proof). Thus $p - 1 = 2^j$ for some j, and $p = 2^j + 1$. Now, j must be a power of 2. Indeed, suppose that $d > 1$ is an odd divisor of j. If $j = de$ and $2^e = a$, then

$$p = 2^j + 1 = 2^{de} + 1 = a^d + 1 = (a+1)(a^{d-1} - a^{d-2} + \cdots - d + 1)$$

which is impossible, since p is prime. Hence $j = 2^i$, and $p = 2^{2^i} + 1 = F_i$ is a Fermat prime.

Next we show that if p is an odd prime, then the regular p^2-gon is never constructible. Indeed, otherwise the degree of the algebraic number $t = 2\cos\frac{2\pi}{p^2}$ would be a power of 2. Then the degree of

$$\eta = \cos\frac{2\pi}{p^2} + i\sin\frac{2\pi}{p^2}$$

would be a power of 2, as well. However, η is a root of

$$\frac{x^{p^2} - 1}{x^p - 1} = x^{p(p-1)} + x^{p(p-2)} + \cdots + 1,$$

and it can be shown that the degree of η is exactly $p(p-1)$. Since $p(p-1)$ cannot be a power of 2, this is a contradiction.

Now suppose that the regular n-gon is constructible. If d is a divisor of n, then the regular d-gon is also constructible, since a suitable subset of the vertices of the n-gon forms a regular d-gon. This implies that if p is an odd prime divisor of n, then the regular p-gon is constructible and thus p must be a Fermat prime. Also, n cannot have a divisor of the form p^2 where p is an odd prime, since the regular p^2-gon is not constructible. Therefore the prime decomposition of n must be of the form $n = 2^k p_1 \cdots p_m$, where $p_1, \ldots, p_m$ are different Fermat primes. This concludes the proof of the "only if" part of Gauss' theorem.

Exercises

3.1. Prove that for $i \geq 2$, F_i is not the sum of two primes.

3.2. Prove that for every $i \geq 2$ the last digit of F_i is 7.

3.3. Prove that if $F_i = p^m$ where p is a prime, then $m = 1$.

3.4. Determine those primes that are smaller than 10^6 and can be written in the form $n^n + 1$ where n is a positive integer.

3.5. Prove that $2^{2^i} + 3$ is composite for infinitely many i. (H)

3.6. Prove that $\sqrt{2} + \sqrt{3}$ is an algebraic number of degree 4.

3.7. Prove that if α is an algebraic number of degree n, then (each value of) $\sqrt{\alpha}$ is an algebraic number of degree at most $2n$. Is it true that the degree of $\sqrt{\alpha}$ is exactly $2n$?

3.8. Prove that if α is an algebraic number of degree n, then α^2 is an algebraic number of degree at most n. Is it true that the degree of α^2 is exactly n?

4

Some Basic Facts About Linear Spaces and Fields

An expression of the form $c_1 x_1 + \cdots + c_n x_n$ is called a *linear combination* of the numbers $x_1, \ldots, x_n$ with coefficients $c_1, \ldots, c_n$.

Let $F \subset \mathbf{C}$ be a field. The set $V \subset \mathbf{C}$ is called a *linear space (or vector space) over F*, if every linear combination of elements of V with coefficients from F belongs to V.

A subset $G \subset V$ is said to be a *generating system of V*, if every element of V can be obtained as a linear combination of elements of G with coefficients from F.

The elements of a subset $H \subset V$ are said to be *linearly independent*, if whenever $x_1, \ldots, x_n$ are different elements of H, $t_1, \ldots, t_n \in F$ and $t_1 x_1 + \cdots + t_n x_n = 0$, then necessarily $t_1 = \cdots = t_n = 0$.

As an illustration of these notions consider the set $V = \mathbf{Q}(\sqrt{2})$. It is clear that V is a linear space over $\mathbf{Q}$. Each of the sets $\{1, 2, \sqrt{2} + 5\}$, $\{1 + \sqrt{2}, 1 - 3\sqrt{2}\}$, $\{1, \sqrt{2}\}$ is a generating system of V. In every linear space any one-element set consisting of a nonzero number is linearly independent. In $\mathbf{Q}(\sqrt{2})$ the set $\{1, \sqrt{2}\}$ consists of linearly independent elements, while $\{1, 2, \sqrt{2} + 5\}$ does not.

The following basic fact is sometimes called the fundamental theorem of linear algebra. We shall accept it without proof.

In every linear space, the cardinality of any generating system is not less than the cardinality of any set of linearly independent elements. (In the applications we shall not need the notion of infinite cardinals, since we shall only use the statement for finite generating systems. The notion of cardinality of infinite sets will be discussed in Section 10.)

If a set is a generating system and, at the same time, it consists of linearly independent elements, then it is called a *basis*. It is easy to see that a set B is a basis of the linear space V over F if and only if every element of V can be written uniquely as the linear combination of elements of B with coefficients from F.

One can prove that *every linear space has a basis*. Moreover, *every set of linearly independent elements is contained in a basis*. It follows from the fundamental theorem that *in every linear space V the cardinality of any two bases is the same*. This common cardinality is called the *dimension* of the linear space V. The space is called *finite dimensional* if its dimension is finite (that is, if it has a finite basis).

Let F be a field. A field K is called an *extension of F* if $F \subset K$. In this case K is a linear space over F (why?). The dimension of this linear space is called the *degree* of the extension $F \subset K$ and is denoted by $[K : F]$. The extension $F \subset K$ is called *finite* if $[K : F]$ is finite.

For example, let $a \in F$ be a positive real number such that $\sqrt{a} \notin F$. Let $K = F(\sqrt{a})$. Then the set $\{1, \sqrt{a}\}$ is a generating system of K and consists of linearly independent elements. Therefore it is a basis, and thus the dimension of K over F is 2. In other words, $F(\sqrt{a})$ is a finite extension of F and $[F(\sqrt{a}) : F] = 2$.

The notion of $F(\sqrt{a})$ can be generalized as follows. Let F be a field and let α be an arbitrary complex number. If a field K contains both F and α, then K must also contain every number of the form

$$\frac{a_n \alpha^n + \cdots + a_1 \alpha + a_0}{b_k \alpha^k + \cdots + b_1 \alpha + b_0} \qquad (a_0, \ldots, a_n, b_0, \ldots, b_k \in F). \qquad (1)$$

It is clear that the set of numbers listed in (1) forms a field and therefore it is the smallest field that contains both F and α. We shall denote this field by $F(\alpha)$. Note that this notation is in accordance with the earlier definition of $F(\sqrt{a})$. That is, if $a \in F$ is a positive real number and $\alpha = \sqrt{a}$, then the field $F(\sqrt{a})$ coincides with the new notion of $F(\alpha)$. The fact that $[\mathbf{Q}(\sqrt{2}) : \mathbf{Q}] = 2$ is a special case of the following theorem.

If α is algebraic, then $\mathbf{Q}(\alpha)$ is a finite extension of $\mathbf{Q}$, and $[\mathbf{Q}(\alpha) : \mathbf{Q}]$ equals the degree of α.

Proof. Let n denote the degree of α, and put

$$F = \{r_{n-1} \alpha^{n-1} + \cdots + r_1 \alpha + r_0 : r_i \in \mathbf{Q} \ (i = 0, \ldots, n-1)\}.$$

Then F is a linear space over $\mathbf{Q}$, and the set $\{1, \alpha, \ldots, \alpha^{n-1}\}$ is a generating system of F. These elements are linearly independent over $\mathbf{Q}$, since otherwise there would be a polynomial $q \in \mathbf{Q}[x]$ of degree $\leq n - 1$ such that $q(\alpha) = 0$. This, however, would contradict the fact that the degree of α is n. Therefore F has a basis of n elements; that is, the dimension of F is n.

It is clear that $F \subset \mathbf{Q}(\alpha)$. We shall prove that F is a field. Since $\mathbf{Q}(\alpha)$ is the smallest field containing α, this will prove $\mathbf{Q}(\alpha) = F$ and $[\mathbf{Q}(\alpha) : \mathbf{Q}] = n$.

Obviously, the sum and difference of elements of F also belong to F. In order to prove that the product of two elements of F also belongs to F, it is enough to show that $\alpha^k \in F$ for every $k = 0, 1, \ldots$. We shall prove this by induction on k. The statement is obviously true if $k \leq n - 1$. Let $k \geq n$, and suppose that $1, \alpha, \ldots, \alpha^{k-1} \in F$. Let $p(x) = x^n + c_{n-1}x^{n-1} + \cdots + c_1 x + c_0$ be a polynomial with rational coefficients such that $p(\alpha) = 0$. Then $\alpha^{k-n}p(\alpha) = 0$, and thus

$$\alpha^k = -c_{n-1}\alpha^{k-1} - \cdots - c_1\alpha^{k-n+1} - c_0\alpha^{k-n}.$$

Since $\alpha^i \in F$ for every $i < k$, it follows that the right-hand side belongs to F. Thus $\alpha^k \in F$ for every k, and hence F is closed under multiplication.

Finally, we have to prove that if $\beta \in F$ and $\beta \neq 0$, then $1/\beta \in F$. Every element of F is of the form $f(\alpha)$, where $f \in \mathbf{Q}[x]$ and the degree of f is less than n. (Recall that $\mathbf{Q}[x]$ denotes the set of polynomials with rational coefficients.) Let $\deg f$ denote the degree of the polynomial f. We have to show that if $f \in \mathbf{Q}[x]$, $\deg f < n$, and $f(\alpha) \neq 0$, then $1/f(\alpha) \in F$. We shall prove this statement by induction on $\deg f$. If $\deg f = 0$, then f is a nonzero rational constant, and thus $1/f(\alpha) \in \mathbf{Q} \subset F$.

Suppose that $\deg f = k$ where $0 < k < n$, and that the statement is true for every polynomial $g \in \mathbf{Q}[x]$ of degree $< k$. Applying the division algorithm to the polynomials p and f, we obtain that $p = q \cdot f + r$, where $q, r \in \mathbf{Q}[x]$ and $\deg r < k$. Since $\deg p = n > k$, we have $q \neq 0$. Also, $k > 0$ implies $\deg q < n$. Therefore $q(\alpha) \neq 0$, since otherwise the degree of α would be less than n.

We have $r(\alpha) = p(\alpha) - q(\alpha) \cdot f(\alpha) = -q(\alpha) \cdot f(\alpha) \neq 0$. Then, by the induction hypothesis, $1/r(\alpha) \in F$. Since F is closed under multiplication, this gives

$$1/f(\alpha) = -q(\alpha) \cdot (1/r(\alpha)) \in F,$$

and the proof is complete.

The connection between algebraic numbers and finite extensions of $\mathbf{Q}$ is given by the following theorem: *a number is algebraic if and only if it is contained in a finite extension of* $\mathbf{Q}$.

The "only if" part is immediate from the previous theorem: if α is algebraic, then α is contained in $\mathbf{Q}(\alpha)$ which is a finite extension of $\mathbf{Q}$. To prove the other direction, let F be a finite extension of $\mathbf{Q}$ with $[F : \mathbf{Q}] = n$. Then any system of linearly independent elements of F consists of at most n elements. Hence, for every $\alpha \in F$, the $n + 1$ elements $1, \alpha, \ldots, \alpha^n$ cannot be linearly independent. This means that there are rational numbers $c_0, \ldots, c_n$ such that not all of them are zero and $c_n \alpha^n + \cdots + c_1 \alpha + c_0 = 0$. Then α is algebraic, and this is what we wanted to show.

We shall need the following theorem on the multiplication of degrees.

If $F \subset K$ and $K \subset L$ are finite field extensions, then $F \subset L$ is also a finite extension and $[L : F] = [K : F] \cdot [L : K]$.

Indeed, let $[K : F] = n$ and let $\{a_1, \ldots, a_n\}$ be a basis of K over F. Let $[L : K] = m$ and $\{b_1, \ldots, b_m\}$ be a basis of L over K. Then it is easy to check that

$$\{a_i b_j : i = 1, \ldots, n, \quad j = 1, \ldots, m\}$$

is a basis of L over F, and thus $[L : F] = nm$.

This theorem implies that *if K is a finite extension of $\mathbf{Q}$, then the degree of every element of K divides $[K : \mathbf{Q}]$.*

Indeed, let n be the degree of an element $\alpha \in K$. Then $\mathbf{Q}(\alpha) \subset K$ and $[\mathbf{Q}(\alpha) : \mathbf{Q}] = n$. Since $[K : \mathbf{Q}] = [\mathbf{Q}(\alpha) : \mathbf{Q}] \cdot [K : \mathbf{Q}(\alpha)] = n \cdot [K : \mathbf{Q}(\alpha)]$, we can see that n is a divisor of $[K : \mathbf{Q}]$.

Now we turn to constructible numbers and constructions. We prove Gauss' result used in the previous section: *every constructible number is algebraic and its degree is a power of* 2.

Let t be a constructible number. We proved that there is a sequence of fields $F_0 = \mathbf{Q} \subset \cdots \subset F_m$ such that $t \in F_m$, and for every $k = 1, \ldots, m$ there is a positive number $a_{k-1} \in F_{k-1}$ such that $F_k = F_{k-1}(\sqrt{a_{k-1}})$. Clearly, for every k the degree $[F_k : F_{k-1}]$ equals either 1 or 2 (it is 1 if $F_k = F_{k-1}$). Thus F_m is a finite extension of $\mathbf{Q}$ and its degree equals the product of the degrees $[F_k : F_{k-1}]$ $(k = 1, \ldots, m)$. Consequently, $[F_m : \mathbf{Q}]$ is a power of 2. Therefore $t \in F_m$ is algebraic, and also its degree, as a divisor of $[F_m : \mathbf{Q}]$, is a power of 2.

Exercises

4.1. Prove that 1, $\sqrt{2}$ and $\sqrt[3]{2}$ are linearly independent over $\mathbf{Q}$.

4.2. Prove that if $p_1, \ldots, p_n$ are distinct primes, then 1, $\sqrt{p_1}, \ldots, \sqrt{p_n}$ are linearly independent over $\mathbf{Q}$. (H)

4.3. Let $a_1, \ldots, a_k$ be positive rational numbers. What is the necessary and sufficient condition of the linear independence of $\sqrt{a_1}, \ldots, \sqrt{a_k}$ over $\mathbf{Q}$? (H)

4.4. Let $F \subset K$ be fields and let $V \neq \{0\}$ be a linear space over K. Prove that if the dimension of V as a linear space over F is finite, then K is a finite extension of F. Show that the dimension of V as a linear space over F equals the dimension of V as a linear space over K multiplied by $[K : F]$.

4.5. Let V be a finite dimensional linear space over $\mathbf{Q}$ such that $x, y \in V$ implies $xy \in V$. Prove that V is a field.

4.6. Let α be an algebraic number of degree n. Prove that the degree of α^2 is a divisor of n and the degree of $\sqrt{\alpha}$ is a divisor of $2n$.

4.7. Let t and ε be complex numbers such that $t = \varepsilon + \varepsilon^{-1}$. Prove that if t is an algebraic number of degree n, then ε is also algebraic and its degree divides $2n$. If, in particular, n is a power of 2, then the degree of ε is also a power of two.

5

Algebraic and Transcendental Numbers

In the first section we examined the numbers $\cos r\pi$, where r is rational. We proved that these numbers are irrational except when they equal $0, \pm 1, \pm 1/2$. Now we show that all these numbers are algebraic. Indeed, as we saw in the first section, there are polynomials t_n of degree n with integer coefficients and leading coefficient 1 such that $t_n(2\cos x) = 2\cos nx$. If $r = k/n$ where k, n are integers, then $t_n(2\cos r\pi) = 2\cos k\pi = (-1)^k \cdot 2$, and thus $2\cos r\pi$ is the root of the polynomial $t_n - (-1)^k \cdot 2$. This shows that $2\cos r\pi$ and $\cos r\pi$ are both algebraic numbers of degree at most n. The exact degree is, in fact, smaller than n. One can prove that if $n > 1$ and $\gcd(k, n) = 1$, then the degree of $\cos(k/n)\pi$ is $\phi(n)/2$ if n is odd, and $\phi(n)$ if n is even. Here $\phi(n)$ is the number of positive integers $i < n$ such that $\gcd(i, n) = 1$.

It can be shown that a number is of the form $2\cos r\pi$ $(r \in \mathbf{Q})$ if and only if it is the root of a polynomial p with integer coefficients and leading coefficient 1 such that every root of p is in the interval $[-2, 2]$. (For the "only if" part, see Exercise 5.1.)

The numbers $\cos r\pi$ $(r \in \mathbf{Q})$ can be given by formulas involving only rational numbers, the basic algebraic operations, and radicals. (For short, we shall say that these numbers can be expressed by radicals.) These formulas are relatively simple when the denominator of r is of the form $2^k p_1 \cdots p_m$, where $p_1, \ldots, p_m$ are different Fermat primes; that is, when $\cos r\pi$ is constructible. For example,

$$2\cos\frac{\pi}{4} = \sqrt{2}, \qquad 2\cos\frac{\pi}{5} = \frac{1+\sqrt{5}}{2}, \qquad 2\cos\frac{\pi}{6} = \sqrt{3},$$

$$2\cos\frac{\pi}{8} = \sqrt{2+\sqrt{2}}, \quad 2\cos\frac{\pi}{10} = \sqrt{\frac{5+\sqrt{5}}{2}}, \quad 2\cos\frac{\pi}{12} = \sqrt{2+\sqrt{3}}.$$

In these formulas only square roots of positive real numbers occur. (This is the reason why these numbers are constructible.) The formulas for nonconstructible $\cos r\pi$ are considerably more complicated. They involve radicals of order greater than 2, and the quantities under the radicals are not necessarily positive real numbers. (Since the nth root of a nonzero complex number has n distinct values, we have to specify what we mean by $\sqrt[n]{\alpha}$. A possible convention is that $\sqrt[n]{\alpha}$ is the (unique) complex number β satisfying $\beta^n = \alpha$ and $\arg(\beta) \in (-\pi/n, \pi/n]$.) For example,

$$2\cos\frac{\pi}{7} = \frac{\sqrt[3]{7}}{3}\left[\sqrt[3]{\frac{-1+3\sqrt{-3}}{2}} + \sqrt[3]{\frac{-1-3\sqrt{-3}}{2}}\right] + \frac{1}{3}$$

and

$$2\cos\frac{\pi}{9} = \sqrt[3]{\frac{1+\sqrt{-3}}{2}} + \sqrt[3]{\frac{1-\sqrt{-3}}{2}}.$$

A number defined by a formula involving only rational numbers, basic algebraic operations, and radicals is always algebraic. This follows from the following two statements.

1. If α, β are algebraic numbers, then so are $\alpha \pm \beta$, $\alpha\beta$ and, if $\beta \neq 0$, α/β. (That is, the set of algebraic numbers is a field.)

2. If α is algebraic and n is a positive integer, then $\sqrt[n]{\alpha}$ is also algebraic.

Proof. We say that a number β is *algebraic over a field F*, if β is the root of a nonzero polynomial with coefficients belonging to F. If β is algebraic over F, then $F(\beta)$ is a finite extension of F (this can be proved in the same way as in the case of $F = \mathbf{Q}$).

Let α be an algebraic number. If β is algebraic, then it is also algebraic over $\mathbf{Q}(\alpha)$. Therefore $F = \mathbf{Q}(\alpha)(\beta)$ is a finite extension $\mathbf{Q}(\alpha)$. Since $\mathbf{Q}(\alpha)$ is a finite extension of $\mathbf{Q}$, this implies that F is a finite extension of $\mathbf{Q}$ as well. Thus every element of F is algebraic. Since $\alpha, \beta \in F$, we have $\alpha \pm \beta, \alpha\beta, \alpha/\beta \in F$, and hence all these numbers are algebraic.

The proof of the second statement is much easier. Let α be algebraic, and let $p \in \mathbf{Q}[x]$ be a nonzero polynomial such that $p(\alpha) = 0$. If $\beta = \sqrt[n]{\alpha}$, then $q(\beta) = 0$, where $q(x) = p(x^n) \in \mathbf{Q}[x]$, and thus β is also algebraic.

A classical result of algebra states that every algebraic number of degree 2, 3, or 4 is expressible by radicals. On the other hand, for algebraic numbers of degree 5 this is not true in general. For example, the roots of the polynomial $x^5 - 4x + 2$ cannot be expressed by radicals.

Until now we have not raised the question whether or not *every* number is algebraic. It was J. Liouville who discovered in 1851 that the answer to this question is negative. A number is called *transcendental* if it is not algebraic. Liouville gave explicit constructions for transcendental numbers. Later G. Cantor showed that "most" numbers are transcendental. We shall discuss the discoveries of Liouville and Cantor in the second part of these notes.

The problem of deciding whether a given number is algebraic or not is rather difficult in general. David Hilbert conjectured in 1900 that if α and β are algebraic, $\alpha \neq 0$, 1 and $\beta \in \mathbf{C} \setminus \mathbf{Q}$, then α^β is transcendental. This was proved by A. O. Gelfond in 1934. More exactly, Gelfond proved that under these conditions *each value* of α^β is transcendental. (Since α^β is defined as $e^{\beta \cdot \log \alpha}$ and since $\log \alpha$ has infinitely many values differing from each other in integer multiples of $2\pi i$, it follows that α^β also has infinitely many values.)

Gelfond's theorem implies, for example, that $2^{\sqrt{3}}$ is transcendental. As another application we also show that e^π is transcendental. Indeed, if it were algebraic, then each value of $(e^\pi)^i$ would be transcendental. However, one of the values is $e^{\pi i} = -1$ which is algebraic.

Our next aim is to prove that *e is transcendental.*

Proof. We shall apply an elaborate version of the argument proving the irrationality of π. Suppose that e is algebraic. Then

$$a_n e^n + a_{n-1} e^{n-1} + \cdots + a_0 = 0, \tag{1}$$

where $a_0, a_1, \ldots, a_n$ are *integers* and $a_0 \neq 0$. Let f be a polynomial of degree m. A repeated application of integration by parts ($m + 1$ times) gives

$$\int_0^k f(x) e^{-x} dx = -\left[f(k) + f'(k) + \cdots + f^{(m)}(k) \right] e^{-k}$$
$$+ \left[f(0) + f'(0) + \cdots + f^{(m)}(0) \right].$$

If we multiply this equation by $a_k e^k$, and add the results for $k = 0, 1, \ldots, n$, then, making use of (1), we obtain

$$\sum_{k=0}^n a_k e^k \int_0^k f(x) e^{-x} dx = -\sum_{k=0}^n a_k \left[f(k) + f'(k) + \cdots + f^{(m)}(k) \right]. \tag{2}$$

We shall construct a polynomial f such that the left-hand side of (2) is small, while the right-hand side is a nonzero integer. This will provide the contradiction.

Let N be a positive integer. In the proof of the irrationality of π we showed that if g is a polynomial with integer coefficients and

$$h(x) = \frac{1}{N!}x^N g(x),$$

then $h^{(i)}(0)$ is an integer for every i. This implies that if g is a polynomial with integer coefficients, a is an integer and

$$f(x) = \frac{1}{(N-1)!}(x-a)^N g(x),$$

then $f^{(i)}(a)$ is an integer divisible by N for every i. Indeed, $f(x) = N \cdot h(x-a)$, where $h(x) = x^N g(x+a)/N!$. Since $g(x+a)$ is also a polynomial with integer coefficients, $h^{(i)}(0)$ is an integer, and thus $f^{(i)}(a) = N \cdot h^{(i)}(0)$ is divisible by N for every i.

Now, in what follows, N will be a prime, $N > |a_0| \cdot n$. We put

$$f(x) = \frac{1}{(N-1)!}x^{N-1}(x-1)^N(x-2)^N \ldots (x-n)^N.$$

Then $f^{(i)}(k)$ is an integer divisible by N for every $i = 0, 1, \ldots$ and $k = 1, \ldots, n$. We prove that the numbers $f^{(i)}(0)$ are divisible by N except for $i = N - 1$.

Indeed, $f(x) = \sum_{i=N-1}^{M} c_i x^i / (N-1)!$ where $c_{N-1} = (\pm n!)^N$ and $c_N, \ldots, c_M$ are integers. We have

$$f^{(i)}(0) = \begin{cases} 0 & \text{if } i \leq N-2, \\ (\pm n!)^N & \text{if } i = N-1, \\ c_i N \cdots i & \text{if } i \geq N. \end{cases}$$

Obviously, $N \mid f^{(i)}(0)$ if $i \neq N-1$. On the other hand, $f^{(N-1)}(0) = (\pm n!)^N$ is not divisible by N since N is prime and $N > n$.

Therefore each term of the right-hand side of (2) is divisible by N except for $a_0 f^{(N-1)}(0)$. Then the right-hand side of (2) is a nonzero integer and, consequently, its absolute value is ≥ 1.

On the other hand, if $0 \leq x \leq n$, then

$$|f(x)| \leq \frac{1}{(N-1)!}n^{(n+1)N} = \frac{A^N}{(N-1)!}, \quad \text{where } A = n^{n+1}.$$

Therefore the absolute value of the left-hand side of (2) is at most

$$(n+1) \max \left(|a_0|, |a_1|, \ldots, |a_n|\right) e^n \cdot n \cdot \frac{A^N}{(N-1)!} = C \cdot \frac{A^N}{(N-1)!},$$

where C and A are positive numbers not depending on N. Let $D = CA^2$. If $N - 1 > 2D^2$, then the product $1 \cdot 2 \cdots (N-1)$ contains more than $(N-1)/2$ factors greater than D^2. Thus

$$(N-1)! > \left(D^2\right)^{(N-1)/2} = D^{(N-1)} > C \cdot A^N.$$

For such an N the absolute value of the left-hand side of (2) is less than 1, while the right-hand side is a nonzero integer. This contradiction completes the proof.

It was proved by F. Lindemann in 1882 that e^α is transcendental if α is algebraic and $\alpha \neq 0$. (With $\alpha = 1$ this immediately gives the transcendence of e.) This also implies that π is transcendental. Indeed, if π were algebraic, then so would be πi and hence $e^{\pi i}$ would be transcendental. However, $e^{\pi i} = -1$ is algebraic, a contradiction.

It follows that $\sqrt{\pi}$ is transcendental as well. Since $\sqrt{\pi}$ is the side of the square having the same area as the unit circle, we can see that "squaring the circle" is not possible using ruler and compass alone. This solves the third classical problem of geometric constructions.

We remarked earlier that e^π is transcendental. On the other hand, it is not known whether or not the numbers π^e, πe, $\pi + e$ are transcendental (or even irrational).

Exercises

5.1. Let t_n be the polynomial satisfying $t_n(2 \cos x) = 2 \cos nx$ $(x \in \mathbf{R})$.

a. Prove that, for $n \geq 2$, $2 \cos(k\pi/n)$ is a multiple root of $t_n - (-1)^k \cdot 2$ for every $k = 1, \ldots, n-1$. (H)

b. Prove that

$$t_n^2(x) - 4 = \prod_{k=1}^{2n} \left(x - 2 \cos \frac{k\pi}{n} \right).$$

5.2.   Prove that if n is odd, then
$$\prod_{k=1}^{n-1} \cos \frac{k\pi}{n} = 2^{1-n}.$$

5.3.   Prove that $t_n(x + x^{-1}) = x^n + x^{-n}$ for every $x \neq 0$.

5.4.   Prove that for $n \geq 1$
$$t_n(x) = x^n - nx^{n-2} + \frac{n(n-3)}{2!}x^{n-4} - \frac{n(n-4)(n-5)}{3!}x^{n-6}$$
$$+ \frac{n(n-5)(n-6)(n-7)}{4!}x^{n-8} - \cdots.$$

5.5.   What is the degree of the algebraic number $\tan 1°$? (H)

5.6.   Express $\cos 12° = \cos(\pi/15)$ by radicals.

5.7.   Express $\cos 3° = \cos(\pi/60)$ by radicals.

5.8.   Express $\cos 1° = \cos(\pi/180)$ by radicals.

5.9.   Let α be a nonzero algebraic real number. Prove that $e^{\alpha\pi}$ is transcendental.

5.10. Prove that a number is algebraic if and only if its real and imaginary parts are both algebraic.

5.11. Prove that $\dfrac{e + \pi i}{\sqrt{e + \pi}}$ is transcendental.

6

Cauchy's Functional Equation

The equation $f(x+y) = f(x) + f(y)$ is called *Cauchy's functional equation*. If a real function $f : \mathbf{R} \to \mathbf{R}$ satisfies this equation for every $x, y \in \mathbf{R}$, then f is said to be *additive*. The additivity of a function f means that it commutes with the operation of addition: applying addition first and then the function f results in the same value as applying the function first and then addition. (Algebraists call such a function a *homomorphism* and, if the function maps the set into itself, then it is called an *endomorphism*. An additive function is an endomorphism of the additive group of the reals.)

If f is an additive function, then it follows by induction that

$$f(x_1 + \cdots + x_n) = f(x_1) + \cdots + f(x_n)$$

for every $x_1, \ldots, x_n \in \mathbf{R}$. In particular, $f(nx) = n \cdot f(x)$ for every $n = 1, 2, \ldots$ and $x \in \mathbf{R}$. This is true for $n = 0$ as well, since $f(0 + 0) = f(0) + f(0)$ implies $f(0) = 0$. Therefore $f(-x) = -f(x)$ (that is, f is an odd function), since $f(x) + f(-x) = f(x + (-x)) = f(0) = 0$.

Every linear function $c \cdot x$ is additive. We prove the converse under some mild conditions on the function f.

Let f be an additive function, and suppose that there is an interval $[a, b]$ on which f is bounded either from above or from below. Then f is linear.

Proof. Suppose that f is bounded from above in $[a, b]$. (The proof is similar if f is bounded from below in $[a, b]$.) Let $c = f(1)$; we prove that $f(x) = cx$ for every x. The function $g(x) = f(x) - cx$ is also additive and vanishes at 1. This implies that $g(r) = 0$ for every rational r. Indeed, if n is a positive integer, then $0 = g(1) = g\left(n \cdot \frac{1}{n}\right) = n \cdot g\left(\frac{1}{n}\right)$ and thus $g(1/n) = 0$. Therefore $g(k/n) = k \cdot g(1/n) = 0$ for every $k = 1, 2, \ldots$; that is, g vanishes at the positive rational numbers. Since g is odd, it follows that $g(r) = 0$ for every $r \in \mathbf{Q}$.

Hence, g is periodic mod every rational number. Indeed, if $r \in \mathbf{Q}$, then $g(x + r) = g(x) + g(r) = g(x)$ for every x.

Suppose that $f(x) \leq K$ for every $x \in [a, b]$. Then

$$g(x) = f(x) - cx \leq K + |c| \cdot \max(|a|, |b|) = M$$

if $x \in [a, b]$. But this implies that $g(x) \leq M$ everywhere. Indeed, for every x there is a rational number r such that $x + r \in [a, b]$ (choose $r \in \mathbf{Q}$ from the interval $[a - x, b - x]$), and thus $g(x) = g(x + r) \leq M$, since $x + r \in [a, b]$.

Now $g(nx) = n \cdot g(x)$ implies that $n \cdot g(x) \leq M$ for every x and for every $n = 1, 2, \ldots$, which is possible only if $g(x) \leq 0$. Therefore $g \leq 0$ everywhere. Since g is odd, $-g(x) = g(-x) \leq 0$ for every x, which gives $g \geq 0$ everywhere. Thus $g \equiv 0$ and $f(x) \equiv c \cdot x$.

In light of the previous theorem it is natural to ask whether or not *every additive function is necessarily linear.* We show next that the answer is negative. In fact, we shall prove that *if α and β are nonzero real numbers such that α/β is irrational, then there is an additive function f such that $f(\alpha) \neq 0$ and $f(\beta) = 0$.* (It is clear that such a function cannot be linear.)

The proof is based on the notion of a *Hamel basis.* The set of real numbers is a linear space over $\mathbf{Q}$. A basis of this linear space is called a Hamel basis. That is, a set $H \subset \mathbf{R}$ is a Hamel basis if and only if every real number x has a unique representation of the form

$$x = r_1 h_1 + \cdots + r_n h_n, \quad \text{where } r_1, \ldots, r_n \in \mathbf{Q} \text{ and } h_1, \ldots, h_n \in H. \quad (1)$$

(Two representations are considered identical if they only differ in terms of the form $0 \cdot h_i$.)

As we remarked in Section 4, in every linear space any set of linearly independent elements is contained in a basis. If $\alpha/\beta \notin \mathbf{Q}$, then the numbers α and β are linearly independent over the rationals. Therefore, there is a Hamel basis H that contains α and β. Now let $f(x)$ denote the coefficient of α in the representation (1). (If α does not occur among $h_1, \ldots, h_n$, then let $f(x) = 0$.) Since the representation (1) is unique, the function f is well defined.

Let x and y be arbitrary real numbers. Suppose that the representations of x and y are given by (1) and $y = s_1 h_1 + \cdots + s_n h_n$, respectively. (We may assume that the two representations involve the same elements $h_1, \ldots, h_n \in H$, since otherwise we may enlarge the representations with suitable terms of the form $0 \cdot h_i$.) Then

$$x + y = (r_1 + s_1) \cdot h_1 + \cdots + (r_n + s_n) \cdot h_n.$$

Since $r_i + s_i \in \mathbf{Q}$ for every i, this is the (unique) representation of $x + y$. If $\alpha = h_i$, then $f(x + y) = r_i + s_i = f(x) + f(y)$. If α does not occur among $h_1, \ldots, h_n$, then we have $f(x) = f(y) = f(x + y) = 0$; therefore $f(x + y) = f(x) + f(y)$ holds in both cases. Hence, f is additive.

Since $\alpha, \beta \in H$, the (unique) representations of α and β are $\alpha = 1 \cdot \alpha$ and $\beta = 1 \cdot \beta$. Thus, $f(\alpha) = 1$ and $f(\beta) = 0$, which completes the proof of the theorem.

In the sequel we shall give two applications of Cauchy's functional equation. First we determine those functions that commute with the operations of addition and multiplication. We prove the following.

If a function $f : \mathbf{R} \to \mathbf{R}$ satisfies $f(x + y) = f(x) + f(y)$ and $f(xy) = f(x)f(y)$ for every $x, y \in \mathbf{R}$, then either $f \equiv 0$ or $f(x) \equiv x$.

Indeed, if $x \geq 0$, then $f(x) = f(\sqrt{x} \cdot \sqrt{x}) = f(\sqrt{x})^2 \geq 0$, and thus f is bounded from below in $[0, 1]$. Since f is additive, it follows that $f(x) = c \cdot x$ everywhere with a constant c. Then $f(1) = f^2(1)$ gives $c = c^2$, and thus either $c = 0$ or $c = 1$.

We note that the analogous statement for the set of complex numbers is far from being true. Let F and K be fields. The function $f : F \to K$ is said to be an *isomorphism* between F and K if f is a bijection from F onto K, and if f commutes with addition and multiplication. By an isomorphism of F we mean an isomorphism between F and an arbitrary subfield of $\mathbf{C}$.

It is easy to see that if $F = \mathbf{Q}$, then the only isomorphism of F is the identity. On the other hand, an important theorem of algebra states that for every field F other than $\mathbf{Q}$ there exists an isomorphism of F that is not the identity. More precisely, if F is a finite extension of $\mathbf{Q}$, then the number of isomorphisms of F is exactly $[F : \mathbf{Q}]$. If F is not a finite extension of $\mathbf{Q}$, then there are infinitely many isomorphisms on F. Since $\mathbf{R}$ is not a finite extension of $\mathbf{Q}$ (why?), there are infinitely many isomorphisms of $\mathbf{R}$. However, as we proved above, the ranges of these isomorphisms cannot be subfields of $\mathbf{R}$. The field of complex numbers also has infinitely many isomorphisms. One of these is the complex conjugation that maps each complex number $z = x + yi$ ($x, y \in \mathbf{R}$) into its conjugate $\bar{z} = x - yi$.

The next application of Cauchy's functional equation concerns the area of polygons. Let $t(A)$ denote the area of the polygon A. Then

(i) $t(A) \geq 0$ for every polygon A;
(ii) If A and B are congruent polygons, then $t(A) = t(B)$;
(iii) If A is the union of the nonoverlapping polygons $A_1, \ldots, A_n$, then
$t(A) = t(A_1) + \cdots + t(A_n)$;
(iv) If Q is a square of side length 1, then $t(Q) = 1$.

These properties (i)–(iv) are usually expressed by saying that the area is nonnegative, invariant, additive and normed. Let ϕ be a real-valued function defined on the set of polygons. We prove that *if ϕ is nonnegative, invariant, additive and normed, then necessarily $\phi(A) = t(A)$ for every polygon A.*

Proof. Let $f(x, y) = \phi(R)$, where R is a rectangle of size $x \times y$. By invariance, $\phi(R)$ depends only on the length of the sides of R and thus the definition of f makes sense. Since any rectangle of size $(x_1 + x_2) \times y$ can be decomposed into two rectangles of size $x_1 \times y$ and $x_2 \times y$, the additivity of ϕ gives

$$f(x_1 + x_2, y) = f(x_1, y) + f(x_2, y) \quad (x_1, x_2, y > 0). \tag{2}$$

We obtain in the same way

$$f(x, y_1 + y_2) = f(x, y_1) + f(x, y_2) \quad (x, y_1, y_2 > 0). \tag{3}$$

Let y be fixed. By (2), the function $x \mapsto f(x, y)$ can be extended to $\mathbf{R}$ as an additive function (see Exercise 6.3.a). Since f is nonnegative, this function must be linear; that is $f(x, y) = c(y) \cdot x$ for every x, where the constant $c(y)$ may depend on y but not on x. Clearly, $c(y) \geq 0$ for every $y > 0$. Applying (3) with $x = 1$ gives $c(y_1 + y_2) = c(y_1) + c(y_2)$ for every $y_1, y_2 > 0$. The previous argument shows that the function $c(y)$ must be also linear. If $c(y) = c \cdot y$, then $f(x, y) = cxy$. Since ϕ is normed, $c = 1$ and $f(x, y) = xy$. Therefore $\phi(R) = t(R)$ holds for every rectangle R.

As Figure 2 shows, every triangle is equidecomposable to a rectangle, and thus $\phi(T) = t(T)$ for every triangle T. Now we show that every polygon P can be decomposed into nonoverlapping triangles. If P is convex, then such a decomposition is obtained by taking the diagonals of P connecting one of the vertices with the others. If P is not convex, then we can decompose it into convex polygons; the straight lines that are continuations of the sides of P divide P into nonoverlapping convex polygons. Decomposing these

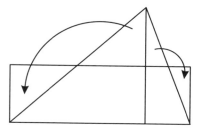

FIGURE 2

convex parts into nonoverlapping triangles we obtain a dissection of P into nonoverlapping triangles.

By the additivity of both ϕ and t this implies that $\phi(P) = t(P)$, which completes the proof.

Exercises

6.1. Let $f : \mathbf{R} \to \mathbf{R}$ be a function such that

$$f\left(\frac{x+y}{2}\right) = \frac{f(x) + f(y)}{2}$$

for every $x, y \in \mathbf{R}$. Prove that there is a constant c such that $f - c$ is additive.

6.2. Let f, g, h be real functions defined on $\mathbf{R}$ such that $f(x + y) = g(x) + h(y)$ for every $x, y \in \mathbf{R}$. Prove that there is a constant c such that $f - c$ is additive.

6.3. **a.** Let $f : (0, \infty) \to \mathbf{R}$ be such that $f(x+y) = f(x) + f(y)$ for every $x, y > 0$. Prove that f can be extended to $\mathbf{R}$ as an additive function.

b. Let $a > 0$ and $f : (0, a) \to \mathbf{R}$ be such that $f(x+y) = f(x) + f(y)$ for every x, $y > 0$, $x + y < a$. Prove that f can be extended to $\mathbf{R}$ as an additive function.

6.4. Let H be a Hamel basis, and let $f : H \to \mathbf{R}$ be an arbitrary function. Prove that f can be extended uniquely to $\mathbf{R}$ as an additive function.

6.5. Suppose that $f : \mathbf{R} \to \mathbf{R}$ satisfies $|f(x + y) - f(x) - f(y)| \leq 1$ for every $x, y \in \mathbf{R}$. Prove that there exists an additive function g such that $|f - g| \leq 1$ everywhere. (H)

6.6. Let $f : \mathbf{R} \to \mathbf{R}$ be an additive function. Prove that if f is not linear, then the graph of f is everywhere dense in the plane. (This means that every rectangle in the plane contains at least one point of the graph of f.)

6.7. Let F be a field and let $f : F \to \mathbf{C}$ be a function commuting with addition and multiplication. Prove that either $f \equiv 0$ or f is an isomorphism between F and $f(F)$.

6.8. Suppose that the function $f : \mathbf{C} \to \mathbf{C}$ commutes with addition and multiplication. Prove that if f is bounded on any subinterval of the real line, then either $f \equiv 0$, or $f(z) \equiv z$, or $f(z) \equiv \overline{z}$.

6.9. Prove that if f is an isomorphism of $\mathbf{Q}(\sqrt{2})$, then either $f(x) \equiv x$ $\left(x \in \mathbf{Q}(\sqrt{2})\right)$ or $f(r + s\sqrt{2}) = r - s\sqrt{2}$ $(r, s \in \mathbf{Q})$.

6.10. Determine the isomorphisms of the field $\mathbf{Q}(\sqrt[3]{2})$.

6.11. Let f and g be additive functions. Prove that the function

$$\phi\big([a, b] \times [c, d]\big) = f(b - a) \cdot g(d - c)$$

is additive on the set of rectangles parallel to the coordinate axes. (H)

7

Geometric Decompositions

In the last section we used the notion of equidecomposability of polygons; now we give a formal definition. The sets $A \subset \mathbf{R}^2$ and $B \subset \mathbf{R}^2$ are called *congruent* (denoted by $A \simeq B$) if there is an isometry (that is, a distance preserving transformation) that maps A onto B.

We say that the polygons A and B are *equidecomposable in the geometric sense*, and write $A \stackrel{g}{\equiv} B$, if there are finite systems of nonoverlapping polygons $A_1, \ldots, A_n$ and $B_1, \ldots, B_n$ such that

$$A = \bigcup_{i=1}^{n} A_i, \quad B = \bigcup_{i=1}^{n} B_i, \quad \text{and} \quad A_i \simeq B_i \text{ for every } i = 1, \ldots, n. \quad (1)$$

It is clear that if $A \stackrel{g}{\equiv} B$, then $t(A) = t(B)$. A formal proof is: supposing (1), the additivity and the invariance of the area imply

$$t(A) = \sum_{i=1}^{n} t(A_i) = \sum_{i=1}^{n} t(B_i) = t(B).$$

The celebrated Bolyai–Gerwien theorem states that the converse is also true: *if A and B are polygons of the same area, then $A \stackrel{g}{\equiv} B$.*

We shall prove the theorem in three steps.

I. First we prove that $\stackrel{g}{\equiv}$ is an equivalence-relation. The properties of reflexivity ($A \stackrel{g}{\equiv} A$) and symmetry ($A \stackrel{g}{\equiv} B \implies B \stackrel{g}{\equiv} A$) are obvious. To prove transitivity, let $A \stackrel{g}{\equiv} B$ and $B \stackrel{g}{\equiv} C$. Then there are systems of nonoverlapping polygons A_i and B_i ($i = 1, \ldots, n$) such that (1) holds. Similarly, there are

39

systems of nonoverlapping polygons B'_j and C_j $(j = 1, \ldots, k)$ such that

$$B = \bigcup_{j=1}^{k} B'_j, \quad C = \bigcup_{j=1}^{k} C_j, \quad \text{and} \quad B'_j \simeq C_j \text{ for every } j = 1, \ldots, k.$$

We may assume that the polygons B_i and B'_j are convex, since otherwise we can decompose them into convex polygons.

Since $A_i \simeq B_i$, there is an isometry ϕ_i of $\mathbf{R}^2$ that maps A_i onto B_i $(i = 1, \ldots, n)$. Similarly, there are isometries ψ_j mapping B'_j onto C_j $(j = 1, \ldots, k)$. For every i, $\bigcup_{j=1}^{k} (B_i \cap B'_j)$ is a decomposition of B_i into nonoverlapping sets, and thus $\bigcup_{j=1}^{k} \phi_i^{-1} (B_i \cap B'_j)$ is a decomposition of A_i into nonoverlapping sets. Similarly, $\bigcup_{i=1}^{n} \psi_j (B_i \cap B'_j)$ is a decomposition of C_j into nonoverlapping sets. Therefore

$$\bigcup_{i=1}^{n} \bigcup_{j=1}^{k} \phi_i^{-1} (B_i \cap B'_j) \quad \text{and} \quad \bigcup_{i=1}^{n} \bigcup_{j=1}^{k} \psi_j (B_i \cap B'_j) \tag{2}$$

are decompositions of A and C into nonoverlapping sets. The sets

$$\phi_i^{-1} (B_i \cap B'_j) \quad \text{and} \quad \psi_j (B_i \cap B'_j)$$

are congruent, since the composition of ϕ_i and ψ_j (which is also an isometry) maps the first set onto the second. This almost proves $A \overset{g}{\equiv} C$; the only snag is that the sets $B_i \cap B'_j$ are not necessarily polygons. However, since B_i and B'_j are convex polygons, it follows that if $B_i \cap B'_j$ is not a polygon, then it is either a closed segment or a single point or it is empty. If we delete these degenerate figures from the unions listed in (2), the remaining terms provide decompositions of A and B into nonoverlapping polygons. This completes the proof of $A \overset{g}{\equiv} C$, and that of the transitivity of $\overset{g}{\equiv}$.

II. We show that *any two rectangles of the same area are equidecomposable using translations alone.* It is clear that a rectangle of size $a \times b$ is equidecomposable to a rectangle of size $(a/2) \times (2b)$ using two pieces and translations. Repeating this process we can see that any rectangle R is equidecomposable to a rectangle R' whose length is at most twice the width. As Figure 3 shows, such a rectangle is equidecomposable to another one with sides parallel to the coordinate axes. (The $ABCD$ rectangle is equidecomposable to the parallelogram $DEFC$ which, in turn, is equidecomposable to $EGHD$.) Now Figure 4 shows that two rectangles of the same area are equidecomposable supposing that they are parallel to the axes and that their lengths are at most

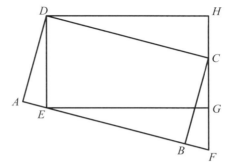

FIGURE 3

twice their widths. Putting these facts together and applying transitivity, we obtain the equidecomposability of arbitrary rectangles of the same area, using only translations.

III. Since, by Figure 2, every triangle is equidecomposable to a rectangle, we obtain that every triangle of area t is equidecomposable to a rectangle of size $1 \times t$. Now let A be an arbitrary polygon of area t. Decomposing A into nonoverlapping triangles, transforming these triangles into rectangles of size $1 \times t_i$ (where $t_1 + \cdots + t_n = t$), and joining these rectangles, we find that A is equidecomposable to a rectangle R of size $1 \times t$. If B is another polygon

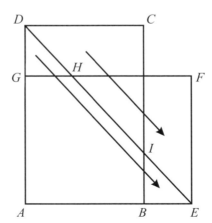

FIGURE 4

of area t, then we obtain $B \overset{g}{\equiv} R$ in the same way, and hence, by transitivity, we have $A \overset{g}{\equiv} B$. This completes the proof of the Bolyai–Gerwien theorem.

In the course of the proof we used only translations and rotations by $180°$. Since the composition of these special isometries are also of this form, it follows that *if A and B are polygons of the same area, then $A \overset{g}{\equiv} B$ using only translations and rotations by* $180°$. This motivates the problem whether or not translations alone would also suffice. Actually, the only place where rotations are needed is the transformation of a triangle into a rectangle (Figure 2). That is, if we could transform a triangle into a rectangle using translations then the answer to this problem would be positive. We show, however, that this is impossible: *a triangle is not equidecomposable to any rectangle using translations alone.*

The idea of the proof is simple. Let F be a function defined on all polygons and suppose that F is additive and invariant under translations. If the polygons A and B are equidecomposable using only translations, then they must satisfy $F(A) = F(B)$. Indeed, if (1) holds, where B_i is a translated copy of A_i for every i, then, by the additivity and translation-invariance of F we obtain

$$F(A) = \sum_{i=1}^{n} F(A_i) = \sum_{i=1}^{n} F(B_i) = F(B).$$

Let Δ be a fixed triangle with vertices A, B, C listed counterclockwise. We shall construct an additive and translation-invariant function F such that $F(\Delta) \neq 0$ and $F(R) = 0$ for any rectangle R. This will prove that Δ is not equidecomposable to any rectangle using translations alone.

Let P be a polygon with vertices $V_0, V_1, \ldots, V_n = V_0$ listed counterclockwise. We define $F(P) = \sum_{i=1}^{n} \varepsilon_i \cdot \overline{P_{i-1}P_i}$, where $\varepsilon_i = 0$ if the side $P_{i-1}P_i$ is not parallel to AB; $\varepsilon_i = 1$ if the side $P_{i-1}P_i$ is parallel to AB and the directed segments $\overrightarrow{P_{i-1}P_i}$ and $\overrightarrow{AB}$ are of the same direction; and $\varepsilon_i = -1$ if the side $P_{i-1}P_i$ is parallel to AB and the directed segments $\overrightarrow{P_{i-1}P_i}$ and $\overrightarrow{AB}$ are of the opposite direction.

It is clear that F is invariant under translations, since a translation does not change the orientation of directed segments. We shall prove that F is additive. Let $A = \bigcup_{i=1}^{n} A_i$, where $A, A_1, \ldots, A_n$ are polygons, and $A_1, \ldots, A_n$ are nonoverlapping. Without loss of generality we may assume that if $i \neq j$, then either $A_i \cap A_j = \emptyset$, or A_i and A_j share either one edge or a vertex. Indeed, otherwise we can decompose A into smaller parts by cutting along the lines of the edges of each A_i. If a segment PQ belongs to the boundary of A_i and lies in the interior of A, then PQ must also belong to the boundary of another A_j with $j \neq i$. Moreover, if we list the vertices of A_i and A_j counterclockwise, then the direction of the segment PQ in these two lists must be opposite to each other. Therefore, when we add up the sums defining $F(A_i)$ $(i = 1, \ldots, n)$, the contributions of this segment will cancel out. After these cancellations, $\sum_{i=1}^{n} F(A_i)$ will only contain the contribution of those segments that form the boundary of A. This proves $F(A) = \sum_{i=1}^{n} F(A_i)$ and thus F is additive.

By the definition of F, we have $F(\Delta) = \overline{AB}$. On the other hand, $F(R) = 0$ for every rectangle R. This is clear if no side of R is parallel to AB. If a side of R is parallel to AB, then the term of $F(R)$ corresponding to this side will be cancelled out by the term corresponding to the opposite side of R. Therefore $F(R) = 0 \neq F(\Delta)$, proving that Δ and R are not equidecomposable with translations.

In the proof of the Bolyai–Gerwien theorem we showed that a rectangle can be transformed into any other rectangle of the same area using translations. We may ask whether the pieces can be chosen to be rectangles in these transformations. In the sequel we shall consider rectangles of the form $[x, y] \times [u, v]$. We shall say that the rectangle R is of size $a \times b$ if $R = [x, y] \times [u, v]$, where $y - x = a$ and $v - u = b$.

Let R and R' be rectangles. We shall write $R \stackrel{r}{\equiv} R'$ if there are nonoverlapping systems of rectangles $R_1, \ldots, R_n$ and $R'_1, \ldots, R'_n$ such that

$$R = R_1 \cup \cdots \cup R_n, \quad R' = R'_1 \cup \cdots \cup R'_n, \quad \text{and} \quad R'_i$$

is a translated copy of R_i for every $i = 1, \ldots, n$. We prove the following.

Let R and R' be rectangles of size $a \times b$ and $c \times d$, and suppose $t(R) = t(R')$. Then $R \stackrel{r}{\equiv} R'$ if and only if $a/c = d/b$ is rational.

For example, the rectangles $[0, 1] \times [0, \sqrt{2}]$ and $[0, \sqrt{2}] \times [0, 1]$ are *not* equidecomposable with translations and rectangular pieces.

The "if" part of the theorem is easy. For, if $a/c = d/b = n/k$ where n, k are positive integers, then $a/n = c/k$ and $d/n = b/k$. Then both R and R' can be decomposed into $n \cdot k$ rectangles of size $(a/n) \times (b/k)$, and this easily implies $R \stackrel{r}{\equiv} R'$.

In order to prove the "only if" part, suppose that $a/c = d/b$ is irrational. Then, as we proved in the last section, there is an additive function f such that $f(a) \neq 0$ and $f(c) = 0$. The function $G([x, y] \times [u, v]) = f(y - x) \cdot (v - u)$

is additive on the set of rectangles with sides parallel to the coordinate axes (see Exercise 6.11). Since G is translation-invariant, it follows that $G(R_1) = G(R_2)$ whenever $R_1 \stackrel{r}{\equiv} R_2$. Now

$$G(R) = f(a) \cdot b \neq 0 \quad \text{and} \quad G(R') = f(c) \cdot d = 0,$$

proving that $R \stackrel{r}{\equiv} R'$ cannot hold.

As an application of the previous theorem we shall prove that *a rectangle can be decomposed into finitely many nonoverlapping squares if and only if the ratio of the sides of the rectangle is rational.*

Proof. The "if" statement is clear: if the ratio of the sides equals n/k, then the rectangle can be decomposed into $n \cdot k$ congruent squares. To prove the converse suppose that R is a rectangle of size $a \times b$, and R is the union of the nonoverlapping squares $Q_1, \ldots, Q_n$. Rotating R by $90°$ we obtain a rectangle R' of size $b \times a$ such that R' is the union of the nonoverlapping squares $Q'_1, \ldots, Q'_n$, where $Q_i \simeq Q'_i$ for every $i = 1, \ldots, n$. Since Q_i can be mapped by a translation onto Q'_i, we have $R \stackrel{r}{\equiv} R'$. Therefore, by our previous theorem, b/a is rational.

The question, whether or not the three dimensional analogue of the Bolyai–Gerwien theorem is true, was raised by Farkas Bolyai (one of the discoverers of the Bolyai–Gerwien theorem) around 1830. Bolyai noted that the proof of the formula giving the area of a triangle (length of base times height, divided by two) is quite simple (see Figure 2). On the other hand, no simple proof of the formula giving the volume of a tetrahedron (area of base times height, divided by three) is known. In fact, all known proofs use the notion of limit in some form. If the Bolyai–Gerwien theorem were true for polyhedra, it could provide an elementary proof. Indeed, if the tetrahedron were equidecomposable to a cube or to a rectangular box of the same volume, then such a decomposition would establish the volume of the tetrahedron in the same way as Figure 2 gives the area of the triangle. Bolyai could not decide this question. Many years later, in 1900, David Hilbert repeated the problem (together with the motivation). Within a year, Max Dehn solved the problem. He proved that *the regular tetrahedron is not equidecomposable to the cube nor to any rectangular box of the same volume.*

The basic idea of the proof of Dehn's theorem is familiar. We shall construct an additive and invariant function Φ defined on all polyhedra such that the value of Φ at a regular tetrahedron is nonzero, while the value of Φ

at any rectangular box is zero. This will imply, in the same way as in the previous cases, that the regular tetrahedron is not equidecomposable to any rectangular box of the same volume.

Let $f : \mathbf{R} \to \mathbf{R}$ be a fixed real function. Let V be a polyhedron, let $e_1, \ldots, e_k$ be the edges of V and let α_i denote the solid angle of V at the edge e_i. We define

$$\Phi(V) = \sum_{i=1}^{k} |e_i| f(\alpha_i),$$

where $|e_i|$ denotes the length of e_i.

It is clear that Φ is invariant under isometries, since the lengths of the edges and the solid angles of a polyhedron remain unchanged under every isometry.

We claim that *if f is additive and if $f(\pi) = 0$, then the function Φ is additive on the set of polyhedra.*

We sketch the proof. Suppose $P = \cup_{i=1}^{n} P_i$, where $P, P_1, \ldots, P_n$ are polyhedra, and $P_1, \ldots, P_n$ are nonoverlapping. We may also assume that if $i \neq j$, then either $P_i \cap P_j = \emptyset$, or P_i and P_j share a face, an edge or a vertex. Indeed, otherwise we can decompose P into smaller parts by cutting along the planes of the faces of each P_i. Suppose that e is an edge of P_i but it is not an edge of P. Then e lies either in the interior of P or in the interior of one of the faces of P. The sum of the solid angles of P_j at the edge e is π if e is in the interior of a face of P, and is 2π if e lies in the interior of P. Since f is additive and $f(\pi) = f(2\pi) = 0$, this easily implies that the contributions of the edge e in the sums defining $\Phi(P_i)$ will cancel out. After these cancellations, $\sum_{i=1}^{n} \Phi(P_i)$ will only contain the contribution of the edges of P. Using the additivity of f it is not difficult to show that the sum of these contributions is exactly $\Phi(P)$. This proves

$$\Phi(P) = \sum_{i=1}^{n} \Phi(P_i),$$

and thus Φ is additive.

Now let A be a regular tetrahedron and B a rectangular box. Since $f(\pi) = 0$, the additivity of f implies $f\left(\frac{\pi}{2}\right) = 0$. As the solid angles of B equal $\frac{\pi}{2}$, this gives $\Phi(B) = 0$.

On the other hand, we have $\Phi(A) = 6df(\alpha)$, where d denotes the common length of the edges of A and α denotes the (common) solid angle of A. If $f(\alpha) \neq 0$, then $\Phi(A) \neq 0 = \Phi(B)$, Consequently, if we can find an f with $f(\alpha) \neq 0$, the proof will be finished.

It is easy to check that $\cos \alpha = \frac{1}{3}$ (see Exercise 7.4.a). This implies that α/π is irrational. Indeed, we proved in the first section that if α/π and $\cos \alpha$ are both rational, then necessarily $\cos \alpha = 0, \pm 1, \pm 1/2$.

Therefore, by one of the theorems of the previous section, there exists an additive function f such that $f(\alpha) \neq 0$ and $f(\pi) = 0$, and this completes the proof.

Exercises

7.1. Prove that Figure 4 is correct. That is, if the rectangles $ABCD$ and $AEFG$ have the same area, then $GHD_\triangle \simeq BEI_\triangle$ and $DIC_\triangle \simeq HEF_\triangle$.

7.2. Prove that a convex polygon A can be decomposed into finitely many nonoverlapping parallelograms if and only if A is centrally symmetric.

7.3. Prove that a convex polygon A is equidecomposable to a square using translations alone, if and only if A is centrally symmetric.

7.4. **a.** Let α be the solid angle of the regular tetrahedron. Prove that $\cos \alpha = 1/3$.

b. Let β denote the solid angle of the octahedron. Prove that $\beta = \pi - \alpha$. In particular, $\cos \beta = -1/3$. (H)

7.5. Prove that the regular tetrahedron and the octahedron are not equidecomposable.

Part II

![Part II banner]

Constructions, Proofs of Existence

Suppose we have to decide whether a mathematical object, construction or procedure exists or not. If the answer is *no*, then, as we saw in the first part, this is final, and solves the problem once and for all. If, on the other hand, the answer is *yes,* this need not be the last word in the solution. This does not mean that there are different degrees of existence. On the contrary: the object in question, as any mathematical object, either exists or does not exist. However, the solution itself can be graded according to how directly it leads to the existence of the given object.

The solution is considered most satisfactory if it consists of a construction that exhibits the given object explicitly. This is called a constructive solution. Unfortunately, in some cases we are not able to give a construction, only a proof that states the existence of the object. These are the so-called proofs of existence. For example, in the next section we shall prove that if a and b are coprime integers and if d is a divisor of $a^2 + b^2$, then d is also the sum of two squares. However, the proof does not produce any representation $d = x^2 + y^2$, it only proves that such a representation exists. (If we know that there is such a representation, then we can find one by checking, for every $x \leq \sqrt{d}$, whether $d - x^2$ is a square or not. Clearly, this should not be considered a construction; besides, it is not feasible if d is very large.) This situation is typical for every existence proof using the *pigeonhole principle* or, more generally, any counting argument.

An important class of existence proofs are those using the axiom of choice (see Section 12). In these proofs the "non-constructive element" is restricted to the application of this axiom; the rest is usually constructive. Ex-

47

amples are provided by the paradoxical decompositions discussed in Sections 12 and 13.

The purest forms of proofs of existence use indirect proof or *reductio ad absurdum*: they suppose that the object in question does not exist and arrive at a contradiction. Most of these proofs, by the very nature of the argument they use, are "pure existence proofs" in that they prove the existence of something without giving the slightest hint how to find it. The following is a typical example of a pure existence proof.

The so-called "divisor game" is played by two players. They choose a natural number n and name distinct, positive divisors of n by turns. The rules are the following.

(i) No divisor of n that is a multiple of a previously pronounced number can be named.

(ii) The player who is forced to name 1 loses the game. The other player is the winner.

Obviously, the game is over in at most $d(n)$ steps (where $d(n)$ denotes the number of positive divisors of n). This implies that, at each point of the game, one of the players has a winning strategy. (Prove it by induction on $k = d(n) - m$, where m is the number of divisors already pronounced.)

We prove that, in fact, *the first player has a winning strategy*.

Suppose this is not true. Then, after each possible first move of the first player, the second player will have a winning strategy. In particular, if the first move of the first player is "n", then the second player has a winning strategy. Now if the first player follows *this* strategy from the beginning, then he/she wins, for the number n can never be named in the course of the game, by the rule (i). But this is impossible, since, by assumption, the first player does not have a winning strategy. This contradiction proves that the first player has a winning strategy.

The argument above proves the existence of a winning strategy without giving any hint how to find one. In fact, no winning strategy has been found (for arbitrary n) to date.

8

The Pigeonhole Principle

Suppose that several objects are distributed among some boxes (for example, letters among pigeonholes) such that the number of the objects is greater than the number of boxes. Then there must be a box that contains at least two objects. This simple observation is called the *pigeonhole principle*, and is used frequently in proofs of existence. In this section we shall give three examples. The first one is combinatorial, and deals with *Sidon sequences*.

A sequence $a_1 < \cdots < a_k$ of positive integers is called a *Sidon sequence* if the numbers $a_i + a_j$ $(1 \leq i \leq j \leq k)$ are all different. Now the problem is the following: what is the length of the longest possible Sidon sequence satisfying $a_k \leq n$?

Consider, for example, $n = 100$. If we want to find a long Sidon sequence, we may try the "greedy algorithm" that chooses at each step the first number that does not violate the Sidon property. In this way we obtain the numbers 1, 2, 4, 8, 13, 21, 31, 45, 66, 81, 97. But this is not the longest Sidon sequence up to 100; for example 1, 3, 7, 25, 30, 41, 44, 56, 69, 76, 77, 86 is a longer one. (This shows that the greedy algorithm need not give the best result.)

Let $s(n)$ denote the length of the longest Sidon sequence satisfying $a_k \leq n$. Then we have $s(100) \geq 12$ and it can be shown by a computer search that, in fact, $s(100) = 12$. The value of $s(n)$ is not known in general. An upper bound can be obtained by using the pigeonhole principle, as follows.

Let $a_1 < \cdots < a_k$ be a Sidon sequence with $a_k \leq n$. There are $\binom{k}{2} + k$ pairs of indices (i, j) satisfying $1 \leq i \leq j \leq k$. Since $2 \leq a_i + a_j \leq 2n$ for every such pair, it follows that $\binom{k}{2} + k \leq 2n - 1$. (Here the objects are the numbers $a_i + a_j$ and the boxes are the possible values $2, \ldots, 2n$. If

$\binom{k}{2} + k > 2n - 1$, that is, the number of objects is greater than the number of boxes, then there would be two objects in the same box, contradicting the Sidon property.) This gives $k(k-1) + 2k \leq 4n - 2$, $k^2 < 4n$, $k < 2\sqrt{n}$ and thus $s(n) < 2\sqrt{n}$.

A better estimate is obtained if we also observe that the numbers $a_j - a_i$ ($i < j$) are different. Since $1 \leq a_j - a_i \leq n-1$ for every $i < j$, the pigeonhole principle gives $\binom{k}{2} \leq n - 1$ and $(k-1)^2 < 2 \cdot \binom{k}{2} \leq 2(n-1) < 2n$; hence $s(n) < \sqrt{2} \cdot \sqrt{n} + 1$.

The best known upper bound is $\sqrt{n} + \sqrt[4]{n} + 1$. Paul Erdős conjectured that $|s(n) - \sqrt{n}|$ is bounded. Erdős considered this problem so important that he offered \$1000 for the proof or disproof of this statement.

The next application is the following theorem: *If a and b are coprime integers, then every positive divisor of $a^2 + b^2$ is the sum of two squares.*

For example, the positive divisors of $10001 = 100^2 + 1^2$ are 1, 73, 137, and 10001 itself. Each of these numbers is the sum of two squares: $1 = 1^2 + 0^2$, $73 = 8^2 + 3^2$, and $137 = 11^2 + 4^2$.

To prove the general statement, let n be a positive divisor of $a^2 + b^2$. We may assume that n is not a perfect square. We shall prove that there are integers x, y, at least one of which is nonzero, such that n divides $x^2 + y^2$, and $|x|, |y| \leq [\sqrt{n}]$. The last condition implies $x^2 + y^2 \leq 2[\sqrt{n}]^2 < 2n$. Since $x^2 + y^2$ is a nonzero multiple of n, this gives $x^2 + y^2 = n$, which is to be proven.

Since $\gcd(a, b) = 1$, it follows that $\gcd(n, b) = 1$ (why?). Hence, n is a divisor of $x^2 + y^2$ if and only if n divides

$$b^2 x^2 + b^2 y^2 = b^2 x^2 - a^2 y^2 + (a^2 + b^2) y^2 = (bx - ay)(bx + ay) + (a^2 + b^2) y^2.$$

Therefore $n \mid x^2 + y^2$ will be satisfied if $n \mid bx - ay$.

Now consider the numbers $bx - ay$, where $0 \leq x, y \leq [\sqrt{n}]$. There are $([\sqrt{n}] + 1)^2 > n$ such numbers. However, the possible remainders of these numbers when divided by n are $0, 1, \ldots, n - 1$, and thus, by the pigeonhole principle, there are distinct pairs (x_1, y_1) and (x_2, y_2) such that $bx_1 - ay_1$ and $bx_2 - ay_2$ give the same remainder. Let $x = x_1 - x_2$ and $y = y_1 - y_2$. Then $bx - ay = (bx_1 - ay_1) - (bx_2 - ay_2)$ is divisible by n. Also, $|x| \leq \max(x_1, x_2) \leq [\sqrt{n}]$, and similarly, $|y| \leq [\sqrt{n}]$. Finally, x and y cannot be both zero, since the pairs (x_1, y_1) and (x_2, y_2) were distinct. This completes the proof.

As an important application, we obtain the following theorem of Fermat: *every prime of the form $4k + 1$ is the sum of two squares.*

Indeed, by a basic theorem of number theory (the so-called Wilson theorem), p divides $(p - 1)! + 1$. If $p = 4k + 1$, then it is easy to check that $(2k)!$ and $(2k+1)(2k+2)\cdots(4k)$ give the same remainder when divided by $4k + 1$. Therefore p divides $[(2k)!]^2 + 1$ and then, by the previous theorem, p is the sum of two squares.

For example, 1997 is a prime of the form $4k + 1$ and, accordingly, it is the sum of two squares: $1997 = 34^2 + 29^2$.

The next application of the pigeonhole principle deals with approximation by rationals. Every number can be approximated by rational numbers with an arbitrary small error. For example, if we want to approximate the number α with an error less than, say, 10^{-6} then we can proceed as follows: we find an integer p such that $p/10^6 \leq \alpha < (p + 1)/10^6$; then the error $|\alpha - (p/10^6)|$ will be smaller than 10^{-6}. Similarly, for every q there is a p such that $|\alpha - (p/q)| < 1/q$.

In general the error $|\alpha - (p/q)|$ is not much smaller than $1/q$. For example, if $\alpha = 1/3$ and if q is not divisible by 3, then for every p we have $|\alpha - (p/q)| = |q - 3p|/(3q) \geq 1/(3q)$. Or, if the decimal expansion of α does not contain the digits 0 and 9, then $|\alpha - (p/q)| \geq 1/(10q)$ whenever q is a power of 10 (why?). However, using other denominators, we can find much more effective approximations.

Consider for example $\alpha = \sqrt{2}$. In the fifth proof of the irrationality of $\sqrt{2}$ we remarked that $(\sqrt{2} - 1)^n = a_n + b_n\sqrt{2}$, where a_n, b_n are integers. Since $(\sqrt{2} - 1)^n \to 0$, $|b_n| \to \infty$ (why?). It follows from the binomial theorem that $(-\sqrt{2} - 1)^n = a_n - b_n\sqrt{2}$, and thus

$$(a_n + b_n\sqrt{2})(a_n - b_n\sqrt{2}) = (-1)^n.$$

Let $p_n = |a_n|$, $q_n = |b_n|$. Then

$$|p_n + q_n\sqrt{2}| \cdot |p_n - q_n\sqrt{2}| = 1,$$

from which $q_n|p_n - q_n\sqrt{2}| < 1$, and

$$\left|\sqrt{2} - \frac{p_n}{q_n}\right| < \frac{1}{q_n^2}.$$

That is, $|\sqrt{2} - (p/q)| < 1/q^2$ holds for infinitely many rationals p/q. Next we show that every irrational number has this property.

For every irrational α there are infinitely many rationals $\frac{p}{q}$ with

$$\left|\alpha - \frac{p}{q}\right| < \frac{1}{q^2}.$$

Proof. Let n be a natural number and consider the numbers $0, \{\alpha\}, \{2\alpha\}, \ldots,$ $\{n\alpha\}$, where $\{x\}$ denotes the fractional part of x; that is $\{x\} = x - [x]$.

These numbers define $n + 1$ points distributed among the n intervals $\left[\frac{i-1}{n}, \frac{i}{n}\right)$ $(i = 1, 2, \ldots, n)$. There must be an interval which contains at least two of these points, and therefore there are numbers $0 \leq k < m \leq n$ such that $|\{m\alpha\} - \{k\alpha\}| < \frac{1}{n}$. Let $q_n = m - k$ and $p_n = [m\alpha] - [k\alpha]$. Then $0 < q_n \leq n$ and $|q_n\alpha - p_n| < 1/n \leq 1/q_n$, from which $|\alpha - (p_n/q_n)| < 1/q_n^2$.

In order to complete the proof we have to check that there are infinitely many different numbers among the quotients p_n/q_n. This is an immediate consequence of $0 < |q_n\alpha - p_n| < 1/n$.

Exercises

8.1. Prove that for every odd integer n there is a positive integer i such that $n \mid 2^i - 1$.

8.2. Let $a_1, a_2, \ldots, a_{100}$ be a sequence of integers. Prove that there is a subsequence $a_{i_1}, a_{i_2}, \ldots, a_{i_k}$ such that 100 divides $a_{i_1} + a_{i_2} + \cdots + a_{i_k}$.

8.3. Let $a_1, \ldots, a_{51}$ be integers with $1 \leq a_i \leq 100$ $(i = 1, \ldots, 51)$. Prove that $a_i \mid a_j$ holds for some $i \neq j$. (H)

8.4. The Fibonacci sequence u_n is defined as follows. Let $u_0 = 0$, $u_1 = 1$, and $u_n = u_{n-1} + u_{n-2}$ $(n = 2, 3, \ldots)$. Prove that there is an $n > 0$ such that $1000 \mid u_n$.

8.5. Let p_n, q_n be as in Exercise 1.3, and let a_n, b_n be integers such that $(\sqrt{2} - 1)^n = a_n + b_n\sqrt{2}$. Prove that $a_n = p_n$ and $b_n = -q_n$.

8.6. Let $r_1 = 1$, $s_1 = 0$, and $r_{n+1} = 2r_n + 3s_n$, $s_{n+1} = r_n + 2s_n$ $(n = 1, 2, \ldots)$. Prove that $|\sqrt{3} - (r_n/s_n)| < 1/s_n^2$. (H)

8.7. **a.** Let D be a positive integer, and let n be a positive divisor of $a^2 + Db^2$, where a and b are coprime integers. Prove that there are integers i, x, y such that $1 \leq i \leq D$ and $i \cdot n = x^2 + Dy^2$.

b. Prove that if $\gcd(a, b) = 1$, then every positive divisor of $a^2 + 2b^2$ is of the form $x^2 + 2y^2$ with integer x, y. Check the result on the divisors of $10002 = 100^2 + 2 \cdot 1^2$.

c. Prove that if $\gcd(a, b) = 1$, then every positive and odd divisor of $a^2 + 3b^2$ is of the form $x^2 + 3y^2$ with integer x, y. Check the result on the divisors of $10003 = 100^2 + 3 \cdot 1^2$.

8.8. Let $\alpha_1, \ldots, \alpha_n$ be irrational numbers. Prove that there are infinitely many n-tuples of rationals $\left(\frac{p_1}{q}, \ldots, \frac{p_n}{q} \right)$ such that

$$\left| \alpha_i - \frac{p_i}{q} \right| < \frac{1}{q^{1+\frac{1}{n}}} \quad (i = 1, \ldots, n).$$

9

Liouville Numbers

Let α and s be real numbers. We shall say that α *is approximable by rationals to order s* if there is a constant c (depending only on α and s) such that

$$\left| \alpha - \frac{p}{q} \right| < \frac{c}{q^s} \tag{1}$$

for infinitely many rational numbers p/q. The last theorem of the previous section states that every irrational number is approximable by rationals to order 2. For some irrationals this is the best we can say. For example,

$$\left| \sqrt{2} - \frac{p}{q} \right| > \frac{1}{4q^2} \tag{2}$$

for every rational $\frac{p}{q}$. Indeed, we have either $|\sqrt{2} - (p/q)| > 1$, or $|\sqrt{2} - (p/q)| \le 1$. In the former case (2) is true. In the latter case we have $0 < p/q < \sqrt{2} + 1$ and

$$\frac{1}{q^2} \le \frac{|2q^2 - p^2|}{q^2} = \left| 2 - \frac{p^2}{q^2} \right| = \left| \sqrt{2} + \frac{p}{q} \right| \cdot \left| \sqrt{2} - \frac{p}{q} \right| < 4 \cdot \left| \sqrt{2} - \frac{p}{q} \right|,$$

which gives (2) again.

Now (2) implies that $\sqrt{2}$ *is not approximable to any order greater than two*. Indeed, let $s > 2$ and suppose that (1) holds with $\alpha = \sqrt{2}$. Then, by (2), $1/4q^2 < c/q^s$, $q^{s-2} < 4c$, and thus $q < (4c)^{1/(s-2)} = d$. The constant d is independent of q and hence any rational number satisfying (1) with $\alpha = \sqrt{2}$ and $s > 2$ has a bounded denominator. If q is fixed, then p is bounded too, and thus the number of rationals p/q satisfying (1) with $\alpha = \sqrt{2}$ and $s > 2$ is finite.

The following theorem generalizes these observations. It was proved by J. Liouville in 1851.

Let α be an algebraic number of degree $n > 1$. Then there is a constant $c > 0$ such that

$$\left| \alpha - \frac{p}{q} \right| \geq \frac{c}{q^n}$$

for every rational $\frac{p}{q}$. Consequently, α is not approximable to any order greater than n.

Proof. Since α is algebraic of degree n, there is a polynomial

$$f(x) = a_n x^n + \cdots + a_1 x + a_0$$

such that $a_0, a_1, \ldots, a_n$ are integers, $a_n \neq 0$ and $f(\alpha) = 0$. Also, the degree of f is the smallest possible, and this implies that f does not have rational roots. Indeed, if r were a rational root of f, then $g(x) = f(x)/(x - r)$ would be a polynomial with rational coefficients. Since $g(\alpha) = 0$ and the degree of g is $n - 1$, this would contradict the minimality of n.

As $f(\alpha) = 0$, it follows that $(x - \alpha)$ is a factor of f, i.e.,

$$f(x) = a_n(x - \alpha)(x^{n-1} + b_{n-2}x^{n-2} + \cdots + b_0).$$

Let p/q be an arbitrary rational number. If $|\alpha - (p/q)| > 1$, then $|\alpha - (p/q)| > 1/q^n$. If $|\alpha - (p/q)| \leq 1$, then $|p/q| \leq |\alpha| + 1 = d$. We have

$$0 \neq \left| f\left(\frac{p}{q} \right) \right| = \left| a_n \frac{p^n}{q^n} + \cdots + a_1 \frac{p}{q} + a_0 \right| = \frac{A}{q^n},$$

where A is a positive integer, and hence $|f(p/q)| \geq 1/q^n$. On the other hand,

$$\left| f\left(\frac{p}{q} \right) \right| = |a_n| \cdot \left| \frac{p}{q} - \alpha \right| \cdot \left| \left(\frac{p}{q} \right)^{n-1} + b_{n-2}\left(\frac{p}{q} \right)^{n-2} + \cdots + b_0 \right|$$

$$\leq \left| \alpha - \frac{p}{q} \right| \cdot |a_n| \left(d^{n-1} + |b_{n-2}|d^{n-2} + \cdots + |b_0| \right)$$

$$= c_1 \cdot \left| \alpha - \frac{p}{q} \right|,$$

where c_1 does not depend on p/q. Therefore $c_1|\alpha - (p/q)| \geq |f(p/q)| \geq 1/q^n$ and thus

$$|\alpha - (p/q)| \geq c/q^n,$$

where $c = \min(1, 1/c_1)$. This proves the first statement of the theorem. The second statement follows in the same way as in the case of $\sqrt{2}$.

A much stronger result was proven by K. F. Roth in 1955. He showed that if α is algebraic, then α is not approximable to any order greater than 2, no matter the degree of α. He did not prove, however, that $|\alpha - (p/q)| > c/q^2$ for every rational number with a constant c. The question whether this stronger statement is also true for every algebraic number is one of the most mysterious unsolved problems of number theory.

An irrational number will be called a *Liouville number* if it can be approximated to any order. From the previous theorem it follows that *every Liouville number is transcendental.*

Since Liouville numbers can be easily constructed, in this way we can produce many transcendental numbers. We show, for example, that

$$\alpha = \sum_{k=0}^{\infty} \frac{1}{10^{k!}}$$

is a Liouville number. Indeed, α is irrational since the decimal expansion of α is not periodic. Let

$$\frac{p_n}{q_n} = \sum_{k=0}^{n} \frac{1}{10^{k!}}.$$

Then $q_n \le 10^{n!}$, and thus

$$\left| \alpha - \frac{p_n}{q_n} \right| = \sum_{k=n+1}^{\infty} \frac{1}{10^{k!}} < \frac{1}{10^{(n+1)!}} \left(1 + \frac{1}{10} + \frac{1}{100} + \cdots \right)$$

$$< \frac{2}{10^{(n+1)!}} \le \frac{2}{q_n^{n+1}} < \frac{1}{q_n^{n}}.$$

Since the numbers p_n/q_n are distinct, this implies that α can be approximated to any order; that is, α is a Liouville number.

It is clear that in this construction 10 can be replaced by any integer greater than 1.

Exercises

9.1. Find all rational numbers satisfying $|\sqrt{2} - (p/q)| < 1/q^3$.

9.2. Find all rational numbers satisfying $|\sqrt[3]{2} - (p/q)| < 1/q^4$.

9.3. Prove that a real number is rational if and only if it is not approximable to any order greater than 1.

9.4. Prove that every rational number is approximable to order 1. (H)

9.5. Prove that an irrational number α is a Liouville number if and only if there is a sequence of rational numbers p_n/q_n such that $q_n > 1$ and $|\alpha - (p_n/q_n)| < 1/q_n^n$ for every $n = 1, 2, \ldots$.

9.6. Prove that

$$\sum_{n=0}^{\infty} 2^{-2^{n^2}}$$

is a Liouville number.

9.7. Let α be a real number, and suppose that p_n/q_n is a sequence of rationals such that $\gcd(p_n, q_n) = 1$, $q_n < q_{n+1} \leq q_n^2$ and $|\alpha - (p_n/q_n)| < c/q_n^2$ for every $n = 1, 2, \ldots$ with a constant $c \geq 1$. Prove that α is not a Liouville number. (H)

9.8. Prove that

$$\sum_{n=0}^{\infty} 2^{-2^n}$$

is not a Liouville number (although it is irrational).

9.9. Prove that if a and b are integers greater than 1, then

$$\sum_{n=0}^{\infty} a^{-b^n}$$

is never a Liouville number (although it is irrational).

10

Countable and Uncountable Sets

Georg Cantor also proved the existence of transcendental numbers—without constructing any of them. His proof is based on the notion of countable sets.

Cantor realized that the set of rational numbers is not "bigger" than the set of natural numbers, in spite of the fact that the rationals are everywhere dense in the real line. In fact, Cantor discovered that the rationals, like the positive integers, can be listed in one single sequence. Such a sequence is, for example:

$$\frac{0}{1}, \frac{-1}{1}, \frac{0}{2}, \frac{1}{1}, \frac{-2}{1}, \frac{-1}{2}, \frac{0}{3}, \frac{1}{2}, \frac{2}{1}, \frac{-3}{1}, \frac{-2}{2}, \frac{-1}{3}, \frac{0}{4}, \frac{1}{3},$$
$$\frac{2}{2}, \frac{3}{1}, \frac{-4}{1}, \frac{-3}{2}, \frac{-2}{3}, \frac{-1}{4}, \frac{0}{5}, \frac{1}{4}, \frac{2}{3}, \frac{3}{2}, \frac{4}{1}, \frac{-5}{1}, \frac{-4}{2}, \ldots \tag{1}$$

Here we listed the quotients p/q according to the value of $|p| + q$; we arranged in some order all quotients with $|p| + q = n$, and combined these finite lists into one single infinite sequence. It is clear that this sequence contains every rational number.

We shall call a set H *countable* if there is an infinite sequence that contains every element of H. A set is *uncountable* if it is not countable. Now the existence of transcendental numbers is an immediate consequence of the following facts:

1. the set of algebraic numbers is countable, and
2. the set of all real numbers is uncountable.

Indeed, these imply that the set of real numbers cannot be the same as the set of algebraic numbers; that is, transcendental numbers must exist.

First we prove that *the set of algebraic numbers is countable*.

Proof. Let us define the weight of a polynomial $a_n x^n + \cdots + a_0$ to be the number $n + |a_n| + |a_{n-1}| + \cdots + |a_0|$. There are only a finite number of polynomials with integer coefficients having a given weight. There is no non-constant polynomial of weight 0 or 1. Arranging the nonconstant polynomials of weight 2 in some order, then those of weight 3, and so on, we obtain a sequence $f_1, f_2, \ldots$ containing every nonconstant polynomial with integer coefficients. Each of these polynomials has at most a finite number of real zeros. Arrange the zeros of f_1 in some order, then those of f_2, and so on. In this way we obtain an enumeration of all algebraic numbers, which completes the proof.

Next we show that *the set of all real numbers is uncountable.*

We give two proofs.

1. We have to prove that there is no sequence containing every real number. In other words, we have to show that if $\{x_n\}$ is an arbitrary sequence of real numbers, then there is a number x not contained in the sequence. The proof will be based on the following simple observation: if I is a closed interval and c is a given real number then there is a closed subinterval of I that does not contain c. This is obvious: choosing two disjoint closed subintervals of I, at least one of them will not contain c.

Let I_1 be a closed interval not containing x_1. Let I_2 be a closed subinterval of I_1 such that $x_2 \notin I_2$. Proceeding inductively, let I_n be a closed subinterval of I_{n-1} such that $x_n \notin I_n$. The nested sequence of closed intervals I_n has a nonempty intersection. If $x \in \bigcap_{n=1}^{\infty} I_n$, then $x \neq x_n$ for every n. That is, x is not an element of the sequence, and this is what we had to prove.

2. Another construction of such an x is the following. Consider the decimal expansions of the numbers $x_1, x_2, \ldots$:

$$x_1 = \pm n_1.a_1^1 a_2^1 \ldots$$
$$x_2 = \pm n_2.a_1^2 a_2^2 \ldots$$
$$\vdots$$

Let $x = 0.b_1 b_2 \ldots$, where $b_i = 5$ if $a_i^i \neq 5$ and $b_i = 4$ if $a_i^i = 5$. It is clear that x is different from each x_n, which completes the second proof.

As we remarked earlier, in this way we prove the existence of transcendental numbers without actually constructing any of them. However, if we combine these arguments, then they turn into a construction. Indeed, the proof of the first theorem produces a well-defined sequence x_n that contains every algebraic number. Then the proof of the second theorem (for definiteness take the second proof), constructs a well-defined number that cannot be algebraic. It is true that this construction is not as simple or direct as taking $\sum_{n=1}^{\infty} 10^{-n!}$. Still, it *is* a construction, and thus Cantor's proof cannot be considered a "pure existence proof."

From the theorems above we can infer not only the existence of transcendental numbers, but also the stronger statement that *the set of transcendental numbers is uncountable.* We first show that *the union of two countable sets is also countable.*

Indeed, if the sequence x_n contains the elements of a countable set A and if the sequence y_n contains the elements of a countable set B, then the sequence $x_1, y_1, x_2, y_2, \ldots$ contains the elements of $A \cup B$, and thus $A \cup B$ is also countable.

Now suppose that the set of transcendental numbers is countable. Then $\mathbf{R}$ would be the union of two countable sets (the set of algebraic numbers and the set of transcendental numbers), and thus $\mathbf{R}$ would be countable. As we proved above, this is not the case, therefore the set of transcendental numbers is uncountable.

A set will be called *countably infinite* if it is countable and infinite. *Every infinite set contains a countably infinite subset.*

Indeed, if A is infinite, then it is nonempty, and we can choose an element $x_1 \in A$. If $x_1, \ldots, x_n \in A$ have been selected, then $A \neq \{x_1, \ldots, x_n\}$ (since otherwise A would be finite), and thus we can choose an element $x_{n+1} \in A \setminus \{x_1, \ldots, x_n\}$. Thus, by induction, we selected the element x_n for every n. The set $\{x_n : n = 1, 2, \ldots\}$ is a countably infinite subset of A.

Exercises

10.1. Let a_n denote the nth term of the sequence (1). Which is the first n for which $a_n = -17/39$?

10.2. Prove that the set of all finite sequences of integers is countable.

10.3. Prove that the set of all finite English texts is countable.

10.4. Prove that the set of all mathematical theorems is countable.

10.5. Prove that any system of pairwise disjoint intervals is countable. (H)

10.6. Prove that any system of pairwise disjoint discs in the plane is countable. (H)

10.7. **a.** Prove that every field of the form $\mathbf{Q}(\alpha)$ is countable.

b. More generally, if F is a countable field and K is a finite extension of F, then K is also countable.

10.8. Prove that if the sets A and B are countable, then so is $A \times B$.

10.9. Prove that the union of countably many countable sets is countable.

The statement that the set of rationals is not bigger than the set of natural numbers can be made precise as follows. The sets A, B will be called *equivalent (or of the same cardinality, or of the same power)*, if there is a one-to-one map of A onto B. If A and B are equivalent, then we shall write $A \sim B$. It is easy to check that $\sim$ is an equivalence-relation.

In the sequel $\mathbf{N}$ will denote the set of positive integers. If a set A is equivalent to $\mathbf{N}$, then A is countable. For, if f is a bijection from $\mathbf{N}$ onto A, then the sequence $f(1)$, $f(2),\ldots$ lists every element of A. The converse is not true, since any finite set is countable but is not equivalent to $\mathbf{N}$. However, *every countably infinite set is equivalent to $\mathbf{N}$. In particular, $\mathbf{Q} \sim \mathbf{N}$.*

To prove the theorem, suppose that the sequence x_n contains every element of the countably infinite set A. Let n_1 be the smallest index for which $x_{n_1} \in A$. If $n_1,\ldots,n_i$ have been selected, then let n_{i+1} be the smallest index for which $n_{i+1} > n_i$, $x_{n_{i+1}} \in A$ and $x_{n_{i+1}}$ is different from the elements $x_{n_1},\ldots,x_{n_i}$. Then the sequence $x_{n_1}, x_{n_2},\ldots$ lists every element of A exactly once, and hence the map $f(i) = x_{n_i}$ ($i \in \mathbf{N}$) is a bijection from $\mathbf{N}$ onto A. Therefore A and $\mathbf{N}$ are equivalent.

Next we prove that *if A is uncountable and $B \subset A$ is countable, then $A \setminus B \sim A$.*

Indeed, $A \setminus B$ is infinite since otherwise $A = B \cup (A \setminus B)$ would be countable. Let C be a countably infinite subset of $A \setminus B$. Then $C \cup B \sim C$, since C and $C \cup B$ are both countably infinite. This easily implies that

$$A \setminus B = \big(A \setminus (B \cup C)\big) \cup C \sim \big(A \setminus (B \cup C)\big) \cup (C \cup B) = A.$$

If a set A is equivalent to $\mathbf{R}$, then we say that *A is of the power of the continuum,* and denote $|A| = c$. The open interval $(-1, 1)$ is of the power of the continuum, since the function $f(x) = x/(1 + |x|)$ is a bijection from $\mathbf{R}$ onto $(-1, 1)$. (The inverse of f is $f^{-1}(x) = x/(1 - |x|)$ ($x \in (-1, 1)$).) Since every open interval is equivalent to $(0, 1)$ (the linear function $(b - a)x + a$

maps $(0, 1)$ onto (a, b)), it follows that every open interval is of the power of the continuum.

Since, by the previous theorem, $[a, b] \sim (a, b)$, it follows that the closed intervals are also of the power of the continuum. The same is true for half-open intervals.

Next we show that *the set of all infinite* 0-1 *sequences is of the power of the continuum.*

Indeed, let A be the set of all infinite 0-1 sequences, and let B denote the set of those 0-1 sequences that contain infinitely many 0's. Then $A \setminus B$ is countable (why?), and hence $A \sim B$. If $\varepsilon = (\varepsilon_1, \varepsilon_2, \ldots)$ is a 0-1 sequence, then let $f(\varepsilon) = \sum_{i=1}^{n} \varepsilon_i 2^{-i}$. It is easy to see that f is a bijection between B and $[0, 1)$, and thus $A \sim B \sim [0, 1)$.

Exercises

10.10. Prove that a set is infinite if and only if it is equivalent to a proper subset of itself.

10.11. Prove that if A is infinite and B is countable, then $A \cup B \sim A$.

10.12. Prove that the set of irrational numbers is of the power of the continuum.

10.13. Prove that the set of transcendental numbers is of the power of the continuum.

10.14. Prove that every half-line is of the power of the continuum.

10.15. Prove that if $A \cap B = \emptyset$, $|A| = c$ and $|B| = c$, then $|A \cup B| = c$. (H)

10.16. Prove that the boundary of a circle is of the power of the continuum.

10.17. Prove that the power set of $\mathbf{N}$ (that is, the set of all subsets of $\mathbf{N}$) is of the power of the continuum. (H)

10.18. Find an explicit bijection between $(0, 1]$ and the set of infinite sequences of positive integers. (H)

10.19. Prove that if $|A| = c$, $|B| = c$, then $|A \times B| = c$. (H)

10.20. Prove that $|\mathbf{R}^n| = c$ for every $n = 1, 2, \ldots$.

10.21. Prove that $\mathbf{R}$ can be decomposed into a continuum number of disjoint subsets of the power of the continuum. (H)

We conclude this long section by introducing the notion of "not greater" for infinite sets. It is the pigeonhole principle that tells us how to define this notion.

When we say that some objects are distributed among some boxes, we really mean that there is a function that maps the set of objects in question into the set of boxes: $f(x) = y$ denotes that the object x is in the box y. Using this language, the pigeonhole principle states that if there are more objects than boxes, then, for every function f mapping the set of objects into the set of boxes, there are objects $x_1 \neq x_2$ such that $f(x_1) = f(x_2)$. That is, in this case there is no injective map from the set of objects into the set of boxes. In other words, if there is an injective map from the set of objects into the set of boxes, then the number of objects is not greater than the number of boxes. It is this statement that we adopt as a definition for infinite sets as well.

Let A, B be arbitrary sets. We say that *the cardinality of A is not greater than that of B if there is an injective map from A into B.* We denote this by $|A| \leq |B|$.

What makes this notion useful is the following important theorem: *if $|A| \leq |B|$ and $|B| \leq |A|$, then $A \sim B$.*

In order to prove this theorem, we have to show that if there is an injective map from A into B and there is an injective map from B into A, then there is a bijection between A and B. This is an immediate consequence of the following theorem due to Cantor, Bernstein, Schröder and Banach.

If $f : A \to B$ and $g : B \to A$ are injective maps, then there are disjoint decompositions $A = A_1 \cup A_2$ and $B = B_1 \cup B_2$ such that $f(A_1) = B_1$ and $g(B_2) = A_2$.

Proof. For every $x \in A$ we form the sequence

$$x, \ g^{-1}(x), \ f^{-1}\big(g^{-1}(x)\big), \ g^{-1}\Big(f^{-1}\big(g^{-1}(x)\big)\Big), \dots . \tag{2}$$

It can happen that the construction of the sequence (2) stops after a finite number of steps. For example, if $x \notin g(B)$, then we cannot even start, and (2) consists of the single term x. Or, if $x \in g(B)$ but $g^{-1}(x) \notin f(A)$, then (2) consists of two terms.

Let A_1 denote the set of all points $x \in A$ for which (2) is finite and contains an odd number of terms. (Thus A_1 contains, among others, all elements x for which $x \notin g(B)$. But it can also happen that $A_1 = \emptyset$.) We put

$$A_2 = A \setminus A_1, \quad B_1 = f(A_1) \quad \text{and} \quad B_2 = g^{-1}(A_2).$$

The proof will be complete if we show that $B_1 \cup B_2 = B$, $B_1 \cap B_2 = \emptyset$ and $A_2 = g(B_2)$.

Let $y \in B$ be arbitrary. If $g(y) \in A_2$, then $y \in g^{-1}(A_2) = B_2$. If $x = g(y) \in A_1$, then y is the second term of the sequence (2). Since (2) consists of an odd number of terms, y cannot be the last term, that is $y \in f(A)$. Let $y = f(x')$. Then x' is the third term of (2), and thus $x' \in A_1$, since the length of the sequence starting from x' is also odd. Then $y \in f(A_1) = B_1$, which proves $B_1 \cup B_2 = B$.

Next suppose $y \in B_1 \cap B_2$. Then $y \in g^{-1}(A_2)$, and thus $x = g(y) \in A_2$. This means that the sequence (2) starting from x is either infinite or consists of an even number of terms. Since $y \in B_1 = f(A_1)$, $y = f(x')$ where $x' \in A_1$. Now x' is the third term of (2), and thus the sequence starting from x' is either infinite or consists of an even number of terms. This contradicts $x' \in A_1$, and hence $B_1 \cap B_2 = \emptyset$.

Finally, let $x \in A_2$. Then (2) contains at least two terms, so that $x = g(y)$ for some $y \in B$. Then $y \in g^{-1}(A_2) = B_2$ and $x \in g(B_2)$. This proves $A_2 = g(B_2)$.

Exercises

10.22. Prove that if the set $A \subset \mathbf{R}^2$ contains a line segment, then $|A| = c$.

10.23. Prove that the surface of a ball is of the power of the continuum.

10.24. Let $A_1, A_2, \ldots$ be countable sets each containing at least two elements. Prove that the set $\{(a_1, a_2, \ldots) : a_1 \in A_1, \ a_2 \in A_2, \ldots\}$ is of the power of the continuum.

10.25. Prove that if $A \subset B \subset C$ and $A \sim C$, then $A \sim B$.

10.26. Prove that $|A| \leq |B|$ if and only if there is a function mapping B onto A.

10.27. Suppose that there is a function mapping the set A onto B, and there is another function mapping B onto A. Prove that $A \sim B$.

10.28. Let A be a square and B a disc. Prove that there are disjoint decompositions $A = A_1 \cup A_2$ and $B = B_1 \cup B_2$ such that A_1 is similar to B_1 and A_2 is similar to B_2. (H)

11

Isometries of $\mathbf{R}^n$

In this section we shall describe the isometries of $\mathbf{R}^n$ for $n = 1, 2, 3$. Recall that an isometry of $\mathbf{R}^n$ is a distance-preserving bijection from $\mathbf{R}^n$ onto itself. The inverse of an isometry α will be denoted by α^{-1}. If α, β are isometries, then $\alpha\beta$ will denote their composition. (That is, $\alpha\beta(x) = \alpha(\beta(x))$ for every $x \in \mathbf{R}^n$.) It is clear that if α, β are isometries, then so are α^{-1} and $\alpha\beta$.

By a *hyperplane of* $\mathbf{R}$, $\mathbf{R}^2$ *or* $\mathbf{R}^3$ we mean a point, a line or a plane, respectively. In general, a hyperplane of $\mathbf{R}^n$ is a translate of a subspace of dimension $n - 1$. If a, b are distinct points of $\mathbf{R}^n$, then the set $\{x \in \mathbf{R}^n : |x - a| = |x - b|\}$ is a hyperplane; it is called the *perpendicular bisector hyperplane* of the segment ab.

Let H be a hyperplane. Then for every point $x \notin H$ there is a unique point x' such that H is the perpendicular bisector of xx'. The map α defined by $\alpha(x) = x'$ for $x \notin H$ and $\alpha(x) = x$ for $x \in H$ is called the *reflection about* H. In the sequel, by a reflection we shall always mean a reflection about a hyperplane. We start with the following general theorem.

Every isometry of $\mathbf{R}^n$ *can be obtained as the composition of at most* $n + 1$ *reflections.*

Proof. We say that the points $x_1, \ldots, x_{n+1}$ are *in general position* if they are not contained in any hyperplane. (For example, $x_1 = (0, \ldots, 0)$, $x_2 = (1, 0, \ldots, 0)$, $x_3 = (0, 1, 0, \ldots, 0), \ldots, x_{n+1} = (0, \ldots, 0, 1)$ are in general position.) The point x is called a *fixed point* of an isometry α if $\alpha(x) = x$. We shall prove that *if* $x_1, \ldots, x_{n+1}$ *are in general position and if* $x_1, \ldots, x_k$ *are fixed points of an isometry* α, *then* α *is the composition of at most* $n + 1 - k$ *reflections.* We shall prove this statement by backward induction, starting with $k = n + 1$ and descending to $k = 0$. Since every isometry satisfies the condition with $k = 0$, this will imply that every isometry is the composition of at most $n + 1$ reflections.

Let $k = n + 1$. We have to show that if each x_i is a fixed point of an isometry α, then α is "the composition of zero reflections." By this we mean that α must be the identity. Suppose this is not true, and let a be a point such that $b = \alpha(a) \neq a$. Since α is an isometry, we have

$$|x_i - a| = |\alpha(x_i) - \alpha(a)| = |x_i - b| \quad (i = 1, \ldots, n + 1),$$

and thus the points x_i lie on the perpendicular bisector of the segment ab. This, however, contradicts the choice of the points x_i, since they are not contained in any hyperplane. Hence α is the identity, which proves the case $k = n + 1$.

Let $1 \leq k \leq n+1$, and suppose that the statement has been proved for k. Let α be an isometry such that $x_1, \ldots, x_{k-1}$ are fixed points of α. We have to show that α is the composition of at most $n+1-(k-1) = n-k+2$ reflections. This is clear if $\alpha(x_k) = x_k$, since in that case we may apply the induction hypothesis, and find that α is the composition of at most $n+1-k < n-k+2$ reflections. Suppose $\alpha(x_k) \neq x_k$, and let H be the perpendicular bisector hyperplane of the segment with endpoints x_k and $\alpha(x_k)$. Let β denote the reflection about H; then $\beta(x_k) = \alpha(x_k)$. Since

$$|x_i - x_k| = |\alpha(x_i) - \alpha(x_k)| = |x_i - \alpha(x_k)|$$

for every $i = 1, \ldots, k - 1$, it follows that $x_1, \ldots, x_{k-1} \in H$. Therefore $\beta(x_i) = x_i = \alpha(x_i)$ for every $i = 1, \ldots, k - 1$. But $\beta(x_i) = \alpha(x_i)$ is also true for $i = k$, and thus the points $x_1, \ldots, x_k$ are fixed points of $\beta^{-1}\alpha$. By the induction hypothesis, $\beta^{-1}\alpha$ is the composition of at most $n+1-k$ reflections. Therefore $\alpha = \beta(\beta^{-1}\alpha)$ is the composition of at most $n + 1 - k + 1 = n + 1 - (k - 1)$ reflections, which completes the proof.

Now we turn to the description of the isometries of $\mathbf{R}$, $\mathbf{R}^2$ and $\mathbf{R}^3$.

I. The isometries of R. Every isometry of $\mathbf{R}$ is the composition of at most two reflections. The composition of the reflections about the points a and b is the translation by the number $2(b - a)$. Indeed, reflecting $x \in \mathbf{R}$ about a we obtain $2a - x$, and reflecting this point about b we arrive at $2b - (2a - x) = x + 2(b - a)$. Therefore, *every isometry of $\mathbf{R}$ is either a translation or a reflection.*

II. The isometries of R^2. Every isometry of R^2 is the composition of at most three reflections. Let r_ℓ denote the reflection about the line ℓ. The composition of two reflections is either a translation or a rotation. More precisely, if ℓ_1 and ℓ_2 are parallel to each other, then $r_{\ell_1} r_{\ell_2}$ is a translation by a vector v such that $v \perp \ell_1$ and $|v|$ is twice the distance between ℓ_1 and ℓ_2.

If ℓ_1 and ℓ_2 intersect at a point P, then $r_{\ell_1} r_{\ell_2}$ is a rotation about P by twice the angle of ℓ_1 and ℓ_2. For the proof of these statements, we refer to Figures 5 and 6.

This also implies that if ℓ_1 and ℓ_2 are parallel to each other and if we translate them simultaneously by the same vector, then the transformation $r_{\ell_1} r_{\ell_2}$ does not change: it remains the same translation. Similarly, if ℓ_1 and ℓ_2 intersect at a point P, then rotating them simultaneously about P, the transformation $r_{\ell_1} r_{\ell_2}$ will not change; namely it will remain the same rotation.

Now we prove that the composition of three reflections is always a *glide reflection,* that is a composition of a reflection r_ℓ and a translation by a vector

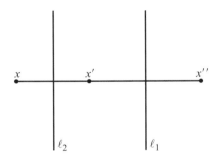

FIGURE 5

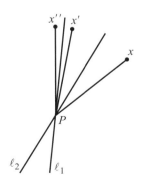

FIGURE 6

parallel to ℓ. (Reflections are considered as special cases of glide reflections when we translate by the zero vector.)

Let $\alpha = r_{\ell_1} r_{\ell_2} r_{\ell_3}$. If $\ell_1 = \ell_2$, then $r_{\ell_1} r_{\ell_2}$ is the identity and $\alpha = r_{\ell_3}$ is a reflection. If $\ell_1 \neq \ell_2$, then, manipulating them simultaneously (translating when they are parallel to each other or rotating otherwise), we may assume that ℓ_2 and ℓ_3 have a common point P. If $\ell_2 = \ell_3$, then $\alpha = r_{\ell_1}$ is, again, a reflection. Otherwise, rotating ℓ_2 and ℓ_3 simultaneously about P, we may assume that ℓ_1 and ℓ_2 are perpendicular to each other. Finally, rotating ℓ_1 and ℓ_2 simultaneously about their intersection, we may assume that ℓ_2 and ℓ_3 are also perpendicular to each other.

We have proved that α is either a reflection or is of the form $r_{\ell_1} r_{\ell_2} r_{\ell_3}$, where ℓ_2 is perpendicular to both ℓ_1 and ℓ_3. It is easy to see that in this case r_{ℓ_1} and r_{ℓ_2} commute; that is $r_{\ell_1} r_{\ell_2} = r_{\ell_2} r_{\ell_1}$. Therefore $\alpha = r_{\ell_2} r_{\ell_1} r_{\ell_3}$. Since $r_{\ell_1} r_{\ell_3}$ is a translation by a vector parallel to ℓ_2, this implies that α is a glide reflection.

Summing up: *Every isometry of $\mathbf{R}^2$ is either a translation or a rotation or a glide reflection.*

Translations and rotations preserve the orientation, while glide reflections change the orientation, of $\mathbf{R}^2$. Therefore, the composition of any number of translations or rotations is again a translation or rotation. Indeed, the composition is an isometry preserving orientation, and hence it must be a translation or a rotation.

III. The isometries of $\mathbf{R}^3$.

Every isometry of $\mathbf{R}^3$ is the composition of at most four reflections. Let r_π denote the reflection about the plane π. The composition of two reflections is either a translation or a rotation about a line. Indeed, if π_1 and π_2 are parallel to each other, then $r_{\pi_1} r_{\pi_2}$ is a translation by a vector v such that $v \perp \pi_1$ and $|v|$ equals twice the distance between π_1 and π_2. If π_1 and π_2 intersect in a line ℓ, then $r_{\pi_1} r_{\pi_2}$ is a rotation about ℓ by twice the angle of π_1 and π_2. These statements can be proven in the same way as in the case of the plane.

Next we consider the compositions of three reflections. Let $\alpha = r_{\pi_1} r_{\pi_2} r_{\pi_3}$. Repeating the argument dealing with three reflections in the plane, we find that either α is a reflection or the planes π_1, π_2, π_3 can be chosen such that π_2 is perpendicular to both π_1 and π_3. Then r_{π_1} and r_{π_2} commute, and thus $\alpha = r_{\pi_2} r_{\pi_1} r_{\pi_3}$. If π_1 and π_3 are parallel to each other, then α is a glide reflection. If π_1 and π_3 intersect, then their intersection is a line perpendicular to π_2. In this case α is a rotation about a line ℓ followed by a reflection about a plane perpendicular to ℓ.

Finally we consider the compositions of four reflections. We prove that such an isometry is necessarily a *helical motion*, that is a rotation about a line ℓ followed by a translation by a vector parallel to ℓ. (Translations and rotations are special cases of helical motions: they correspond to the cases when we translate by the zero vector or rotate by the zero angle.) We shall prove this statement in three steps.

(i) *If α is a rotation about a line ℓ and β is a translation by a vector v perpendicular to ℓ, then $\alpha\beta$ and $\beta\alpha$ are both rotations about lines parallel to* ℓ. Indeed, let $\alpha = r_{\pi_1}r_{\pi_2}$ where π_1, π_2 are planes going through ℓ, and let $\beta = r_{\pi_3}r_{\pi_4}$ where π_3, π_4 are planes perpendicular to v. Then $\alpha\beta = r_{\pi_1}r_{\pi_2}r_{\pi_3}r_{\pi_4}$. Since $v \perp \ell$, the planes π_3, π_4 are parallel to ℓ. Therefore, translating them simultaneously, we may assume that π_3 goes through ℓ. Then, rotating π_1 and π_2 simultaneously about ℓ, we may assume $\pi_3 = \pi_2$. Then $\alpha\beta = r_{\pi_1}r_{\pi_4}$, and thus $\alpha\beta$ is a rotation about the intersection of π_1 and π_4, which is parallel to ℓ. The same argument applies to $\beta\alpha$.

(ii) *If α is a translation and β is a rotation, then $\alpha\beta$ and $\beta\alpha$ are both helical motions.* Indeed, let α be a translation by the vector v, and let β be a rotation about the line ℓ. Let $v = v_1 + v_2$, where v_1 is parallel to ℓ and $v_2 \perp \ell$. If α_i denotes the translation by v_i $(i = 1, 2)$, then, as above, $\alpha_2\beta$ and $\beta\alpha_2$ are both rotations about lines parallel to ℓ. Hence $\alpha\beta = \alpha_1(\alpha_2\beta)$ and $\beta\alpha = (\beta\alpha_2)\alpha_1$ are both helical motions.

(iii) Let $\alpha = r_{\pi_1}r_{\pi_2}r_{\pi_3}r_{\pi_4}$; we prove that α is a helical motion. If π_1 and π_2 are parallel to each other, then $\beta = r_{\pi_1}r_{\pi_2}$ is a translation. Since $\gamma = r_{\pi_3}r_{\pi_4}$ is either a translation or a rotation, it follows that $\alpha = \beta\gamma$ is a helical motion in both cases. Similarly, if π_3 and π_4 are parallel to each other, then α is again a helical motion. Therefore we may assume that π_1 and π_2 intersect each other in a line ℓ_1, and that π_3 and π_4 intersect each other in a line ℓ_2. Rotating π_1 and π_2 simultaneously about ℓ_1, we may assume that π_2 and ℓ_2 have a common point P. Let ℓ_3 be the line going through P and parallel to ℓ_1. Then π_2 contains ℓ_3. Rotating π_3 and π_4 simultaneously about ℓ_2, we may assume that π_3 also contains ℓ_3. Then, rotating π_2 and π_3 simultaneously about ℓ_3, we may assume that π_2 is parallel to π_1, since ℓ_3 and ℓ_1 are parallel to each other. As we saw above, this implies that α is a helical motion, which completes the proof.

Summing up: *The isometries of $\mathbf{R}^3$ are the helical motions (including the translations and rotations as special cases), the glide reflections, and those transformations that are obtained by a rotation followed by a reflection about a plane perpendicular to the axis of the rotation.*

The helical motions preserve the orientation of $\mathbf{R}^3$; the other isometries change its orientation. This implies that the composition of any number of helical motions is again a helical motion.

The following statement will be needed later. *Let α, β be rotations about the lines ℓ_1 and ℓ_2, respectively. If ℓ_1 and ℓ_2 go through the point P, then $\alpha\beta$ is also a rotation about a line going through P.* (In other words, the composition of two rotations of a ball is also a rotation of the same ball.) The proof is immediate: since $\alpha\beta$ is the composition of four reflections, it is a helical motion. On the other hand, P is a fixed point of $\alpha\beta$. It is easy to see that any helical motion having a fixed point P must be a rotation about a line going through P.

Exercises

11.1. Prove that no bounded subset of $\mathbf{R}$ can be congruent to a proper subset of itself.

11.2. Prove that for every set $H \subset \mathbf{R}$ there is at most one point $x \in H$ such that $H \simeq H \setminus \{x\}$.

11.3. Prove that if a bounded set $H \subset \mathbf{R}^2$ is congruent to a proper subset of itself, then the isometry mapping H onto this proper subset is a rotation.

11.4. Suppose that α and β are translations or rotations of $\mathbf{R}^2$. Prove, without referring to the notion of orientation, that $\alpha\beta$ is also a translation or a rotation. (H)

11.5. Prove, without referring to the notion of orientation, that the composition of two helical motions is also a helical motion.

11.6. Prove, without using the notion of helical motion, that the composition of two rotations about intersecting lines is also a rotation. (H)

The aim of the next four exercises is to prove the following theorem: *If f is a bijection from $\mathbf{R}^2$ onto itself such that $|f(x) - f(y)| = 1$ whenever $|x - y| = 1$, then f is an isometry.* (This is a special case of a theorem by F. S. Beckman and D. A. Quarles. They proved this statement in every dimension, even without requiring that f be a bijection.)

11.7. Let f be a bijection from $\mathbf{R}^2$ onto itself such that $|f(x) - f(y)| = 1$ whenever $|x - y| = 1$. Prove that if $x, y \in \mathbf{R}^2$ and $|x - y|$ is an integer, then $|f(x) - f(y)| = |x - y|$. (H)

11.8. Let f be as in the previous exercise. Prove that if $|x - y| = 1/2$, then $|f(x) - f(y)| = 1/2$. (H)

11.9. Let f be as in Exercise 11.7. Prove that if $|x - y| = n/2^k$ where n, k are positive integers, then $|f(x) - f(y)| = n/2^k$.

11.10. Let f be as in Exercise 11.7. (i) Prove that f is continuous. (ii) Prove that f is an isometry. (H)

12

The Problem of Invariant Measures

As we saw earlier, the area is a nonnegative, additive, invariant, and normed function defined on the set of polygons. In the last century several mathematicians considered the problem whether a function with similar properties can be defined on wider classes of sets. Dealing with sets more general than just polygons, it is necessary to modify the notion of additivity. We shall say that a function m is *additive* if $m(A \cup B) = m(A) + m(B)$ holds whenever $A \cap B = \emptyset$, and m is defined on the sets A, B, $A \cup B$. In the sequel, nonnegative and additive functions will be called *measures*. The measure m is called *invariant under a transformation* α if $m(\alpha(A)) = m(A)$ for every set A such that m is defined on A and $\alpha(A)$.

Émile Borel considered those sets which can be obtained from the polygons by a countable number of applications of set theoretic operations. (Now these sets are called *Borel sets*; we shall investigate them in Section 16.) Borel and H. Lebesgue proved that there is a measure m on the family of all Borel sets such that $m(A) = t(A)$ if A is a polygon, and that m is invariant under every isometry. Actually the measure they constructed was not only additive, but also *σ-additive* in the following sense: if $A_1, A_2, \ldots$ are pairwise disjoint sets, then

$$m\left(\bigcup_{n=1}^{\infty} A_n\right) = \sum_{n=1}^{\infty} m(A_n).$$

This motivated the following question: is it possible to find a similar function defined on all subsets of $\mathbf{R}^2$? In other words, does there exist an invariant and σ-additive measure m defined on all subsets of $\mathbf{R}^2$ such that $m(A) = t(A)$ if A is a polygon?

The answer was given by G. Vitali in 1905; he proved that such a function m does not exist. In fact, Vitali proved the following more general theorem.

Let m be a measure defined on all subsets of $\mathbf{R}^n$. If m is invariant under translations and $m(Q_n) = 1$ where Q_n is the cube $[0, 1] \times \cdots \times [0, 1]$, then m cannot be σ-additive.

Although Vitali's theorem claims the *nonexistence* of a certain measure, the proof is based on the *existence* of a set with some strange properties. We shall prove the following.

There exists a set $A \subset \mathbf{R}^n$ such that
(i) *the cube Q_n contains infinitely many disjoint translated copies of A, and*
(ii) *$\mathbf{R}^n$ can be covered by countably many disjoint translated copies of A.*

Suppose for a moment that such a set exists, and let m be a σ-additive and translation-invariant measure defined on all subsets of $\mathbf{R}^n$ such that $m(Q_n) = 1$. Every measure is monotone; that is $B \subset C$ implies $m(B) \le m(C)$. Indeed, $m(C) = m(B) + m(C \setminus B) \ge m(B)$, since $m(C \setminus B) \ge 0$. Let $A_1, A_2 \ldots \subset Q_n$ be disjoint translated copies of A, and let $B = \cup_{k=1}^\infty A_k$. Then $m(B) \le m(Q_n) = 1$ by monotonicity, and $m(B) = \sum_{k=1}^\infty m(A_k)$ by σ-additivity. By translation-invariance we have $m(A_k) = m(A)$ for every k, and this implies $m(A) = 0$. Then, by (ii), we have $m(\mathbf{R}^n) = 0$, and hence $m(Q_n) = 0$ using monotonicity again. This contradiction proves Vitali's theorem.

Now we turn to the construction of the set A. We shall give the construction for $n = 1$; the general case can be treated similarly. If $n = 1$, then the "cube" Q_1 is simply the interval $[0, 1]$.

Let $x \sim y$ if $y - x$ is rational. Then $x \sim x$, $x \sim y$ implies $y \sim x$, and $x \sim y$, $y \sim z$ imply $x \sim z$. Therefore $\sim$ is an equivalence-relation, and thus $\mathbf{R}$ is decomposed into disjoint classes such that $x \sim y$ if and only if x and y belong to the same class. Each class is of the form $\mathbf{Q} + x = \{r + x : r \in \mathbf{Q}\}$ with some fixed x. Hence the classes are everywhere dense in $\mathbf{R}$ and we can pick one point $x \in [0, \frac{1}{2}]$ from each class. Let A denote the set of these points. Then the sets $A + r$ $(r \in \mathbf{Q})$ are pairwise disjoint translated copies of A. Indeed, suppose that r and s are distinct rational numbers, and $(A + r) \cap (A + s) \ne \emptyset$. If $x \in (A + r) \cap (A + s)$, then $x - r, x - s \in A$. However, $x - r \sim x - s$ (since $s - r \in \mathbf{Q}$), and this contradicts the fact that A contains exactly one element from each equivalence class. Now (i) follows from

$$[0, 1] \supset \bigcup_{n=2}^{\infty} \left(A + \frac{1}{n} \right),$$

while (ii) is a consequence of

$$\mathbf{R} = \bigcup_{r \in \mathbf{Q}} (A + r).$$

To prove this equality, note that for every $x \in \mathbf{R}$ there is a $y \in A$ satisfying $y \sim x$. If $x - y = r$, then $r \in \mathbf{Q}$ and $x = y + r \in A + r$. This completes the proof.

In the course of the proof we tacitly assumed that we can select a point from each set of a family of disjoint nonempty sets. This statement was first explicitly formulated by E. Zermelo in 1904. Now it is called the Axiom of Choice, and is one of the standard axioms of set theory.

Motivated by Vitali's theorem, we may ask if there exists an invariant measure m on all subsets of $\mathbf{R}^n$ such that $m(Q_n) = 1$. (We do not require σ-additivity.) For $n = 1$ and $n = 2$ the problem was solved by S. Banach in 1923: Banach proved that measures with these properties do exist. However, as F. Hausdorff showed in 1914, for $n = 3$ there is no such measure. The proof of Hausdorff's theorem is similar to that of Vitali: it is based on the existence of a set with paradoxical properties.

Let S denote the unit sphere in $\mathbf{R}^3$. (That is, $S = \{x \in \mathbf{R}^3 : |x| = 1\}$.) Hausdorff proved the following statement (the so-called "Hausdorff paradox"). *There are decompositions $S = A_1 \cup A_2 \cup C_1$ and $S = A_3 \cup A_4 \cup A_5 \cup C_2$ such that the sets A_i ($i = 1, \ldots, 5$) are congruent and C_1, C_2 are countable.*

This implies that there is no finitely additive isometry-invariant measure m defined on all subsets of S satisfying $m(S) = 1$. Indeed, such a measure vanishes on every countable set $C \subset S$ (see Exercise 12.4). Thus $S = A_1 \cup A_2 \cup C_1$ implies $m(A_1) = 1/2$, while $S = A_3 \cup A_4 \cup A_5 \cup C_2$ implies $m(A_1) = 1/3$, a contradiction.

We shall not prove Hausdorff's paradox here. Instead we shall establish the following theorem, which is easier to prove, and is still sufficient for the applications.

There exists a set $A \subset S$ such that
(i) *S contains infinitely many pairwise disjoint congruent copies of A, and*
(ii) *S can be covered by four congruent copies of A.*

This also implies the nonexistence of invariant and normed measures on S. Indeed, if m were such a measure, then (i) would give $m(A) = 0$, while (ii) would imply $m(A) \geq 1/4$, which is impossible. We shall prove the theorem in five steps.

1. The set of all rotations which map S onto itself is denoted by $SO(3)$. That is, $\phi \in SO(3)$ if ϕ is a rotation about a line going through 0. As was proven in the last section, $\phi, \psi \in SO(3)$ implies $\phi\psi \in SO(3)$. The identity map, denoted by e, is also considered as an element of $SO(3)$. If $\phi \in SO(3)$ is a rotation by an angle α, then ϕ^{-1} denotes the rotation about the same axis by the angle $-\alpha$. Then, for every $\phi, \psi, \chi \in SO(3)$ we have $\phi(\psi\chi) = (\phi\psi)\chi$, $\phi\phi^{-1} = \phi^{-1}\phi = e$, and $e\phi = \phi e = \phi$. That is, $SO(3)$ forms a group.

2. Let ϕ and ψ be fixed elements of $SO(3)$. Consider the set of all possible compositions of ϕ, ϕ^{-1}, ψ, and ψ^{-1}. These are of the form

$$\phi^{a_1}\psi^{a_2}\phi^{a_3}\cdots\psi^{a_{2n}}, \tag{1}$$

where $a_1, \ldots, a_{2n}$ are integers and $a_2, \ldots, a_{2n-1}$, if they exist, are nonzero. We denote here $\phi^0 = \psi^0 = e$ and $\phi^{-n} = (\phi^{-1})^n = (\phi^n)^{-1}$. Let G denote the set of all possible expressions of the form (1). Then G is countable (cf. Exercise 10.2). We consider the empty expression as an element of G; we interpret it as the representation of the identity map e.

Different expressions need not represent different rotations. For example, if ϕ and ψ commute, then $\phi\psi = \psi\phi$, and thus the expressions $\phi^1\psi^1$ and $\phi^0\psi^1\phi^1\psi^0$ represent the same rotation. One can prove, however, that *if ϕ and ψ are rotations about the x-axis and z-axis, respectively, by the same angle α such that $\cos\alpha$ is transcendental, then two different expressions in G represent different rotations.*

In order to prove this lemma one has to show that if two different expressions represent the same rotation, then $\cos\alpha$ satisfies an algebraic equation with integer coefficients. This would require tedious computations; we omit the proof.

3. In the sequel ϕ and ψ will denote fixed rotations as described in the lemma above. We shall identify the expressions of G with the rotations represented by them. We shall say that two points $x, y \in S$ are equivalent, denoted by $x \sim y$, if there is a $\chi \in G$ such that $y = \chi(x)$. For every $x, y, z \in S$ we have $x \sim x$, $x \sim y \implies y \sim x$ and $x \sim y$, $y \sim z \implies x \sim z$; that is, $\sim$ is an equivalence relation. Therefore S is decomposed into disjoint classes such that $x \sim y$ if and only if they belong to the same class. (Each class is the set of all points of S that are equivalent to a given point of S.)

For every $\chi \in G$, $\chi \neq e$, there are exactly two points $x \in S$ such that $\chi(x) = x$. (These are the intersections of the axis of χ with S.) Hence the set

$$C = \{x \in S : \exists \chi \in G, \ \chi \neq e, \ \chi(x) = x\}$$

is countable. We remark that if $x \in C$ and $x \sim y$, then $y \in C$. Indeed, if $\chi_1(x) = x$ and $\chi_2(x) = y$, then

$$\chi_2 \chi_1 \chi_2^{-1}(y) = \chi_2 \chi_1(x) = \chi_2(x) = y,$$

and thus $y \in C$. Therefore C is the union of entire equivalence classes, and then the same is true for $S \setminus C$.

We pick one point of each equivalence class contained in $S \setminus C$, and denote by H the set of these points. (Here we used the Axiom of Choice.) Then $H \subset S \setminus C$, and for every $x \in S \setminus C$ there is exactly one $y \in H$ with $y \sim x$.

Let U denote the set of those expressions (1) in which $a_1 \neq 0$. We define

$$A = \{\chi(x) : \chi \in U, \ x \in H\}.$$

We prove that A satisfies the requirements of the theorem.

4. First we construct infinitely many disjoint congruent copies of A. We claim that the sets $\psi^n(A)$ $(n = 1, 2, \dots)$ are pairwise disjoint. (They are clearly congruent to A.) Suppose this is not true, and let $\psi^n(A) \cap \psi^m(A) \neq \emptyset$, where n, m are distinct positive integers. If $y \in \psi^n(A) \cap \psi^m(A)$, then there are points $x_1, x_2 \in H$ and rotations $\chi_1, \chi_2 \in U$ such that $y = \psi^n \chi_1(x_1) = \psi^m \chi_2(x_2)$. Therefore

$$x_1 = \chi_1^{-1} \psi^{m-n} \chi_2(x_2), \qquad (2)$$

and thus $x_1 \sim x_2$. Since H contains at most one point from each equivalence class, this gives $x_1 = x_2$. Then (2) implies that x_1 is a fixed point of the map $\chi_1^{-1} \psi^{m-n} \chi_2 = \chi$.

Now we claim that χ is not the identity map. Indeed, the expression defining χ when writing it in the form (1) is not the empty expression. This follows from the following facts: $\chi_1, \chi_2 \in U$ implies that χ_1^{-1} ends with a power of ϕ, χ_2 starts with a power of ϕ, and, finally, $n - m \neq 0$. Then, as x_1 is a fixed point of χ, it follows that $x_1 \in C$. This, however, contradicts $x_1 \in H \subset S \setminus C$. This proves that the sets $\psi^n(A)$ are indeed pairwise disjoint.

5. Finally we prove that S can be covered by four congruent copies of A. First we show that

$$S \setminus C \subset A \cup \phi(A). \tag{3}$$

Let $x \in S \setminus C$ be arbitrary. Then there is a point $y \in H$ such that $x \sim y$. Let $x = \chi(y)$, where χ is given by the expression (1). If $a_1 \neq 0$, then $\chi \in U$ and thus $x = \chi(y) \in A$. If $a_1 = 0$, then $\phi^{-1}\chi \in U$. In this case $z = \phi^{-1}\chi(y) \in A$ and $x = \phi(z) \in \phi(A)$. This proves (3).

Since C is countable, there is a rotation $\rho \in SO(3)$ such that

$$C \cap \rho(C) = \emptyset$$

(see Exercise 12.3). Then

$$\rho(C) \subset S \setminus C \subset A \cup \phi(A),$$

and thus

$$C \subset \rho^{-1}(A) \cup \rho^{-1}\phi(A).$$

Therefore

$$S = (S \setminus C) \cup C \subset A \cup \phi(A) \cup \rho^{-1}(A) \cup \rho^{-1}\phi(A).$$

This gives a covering of S with four congruent copies of A, which completes the proof.

Exercises

12.1. Construct a σ-additive measure on all subsets of $\mathbf{R}$ such that the measure of $[0, 1]$ is 1. (H)

12.2. Let m be a σ-additive function defined on all subsets of $\mathbf{R}$ such that $m([0, 1]) = 1$. Prove that m cannot be translation-invariant. (That is, nonnegativity is not needed in Vitali's theorem.) (H)

12.3. Let $C \subset S$ be countable. Prove that there is a rotation $\rho \in SO(3)$ such that $C \cap \rho(C) = \emptyset$.

12.4. Let m be a rotation-invariant measure on all subsets of S such that $m(S) = 1$. Prove that if $C \subset S$ is countable, then $m(C) = 0$. (H)

12.5. Let m be a translation-invariant measure on all subsets of $\mathbf{R}$ such that $m([0, 1]) = 1$. Prove that if $C \subset \mathbf{R}$ is bounded and countable then $m(C) = 0$. (H)

13

The Banach–Tarski Paradox

The concept of equidecomposability of polygons can be generalized to arbitrary subsets of $\mathbf{R}^n$ as follows.

Let $X, Y \subset \mathbf{R}^n$ and suppose that there are disjoint decompositions $X = X_1 \cup \cdots \cup X_k$ and $Y = Y_1 \cup \cdots \cup Y_k$ such that $X_i \simeq Y_i$ $(i = 1, \ldots, k)$. Then we say that X and Y are *equidecomposable,* and write $X \equiv Y$. If we want to indicate that we use k parts in the decompositions, then we write $X \equiv_k Y$.

In Section 7 we proved that the relation $\overset{g}{\equiv}$ (defined for polygons) is an equivalence relation. A similar argument shows that $\equiv$, defined on all subsets of $\mathbf{R}^n$, is also an equivalence relation. The argument actually proves that if $X \equiv_k Y$ and $Y \equiv_m Z$, then $X \equiv_{km} Z$. (The proof is, in fact, easier than in the case of the relation $\overset{g}{\equiv}$, since the parts need not be polygons.)

One can show that if X and Y are polygons in $\mathbf{R}^2$, then $X \overset{g}{\equiv} Y \iff X \equiv Y$ (see Exercise 13.4).

The following theorem is used frequently in problems of equidecomposability.

Suppose $X \equiv_k U \subset Y$ and $Y \equiv_m V \subset X$. Then $X \equiv_{k+m} Y$.

Proof. Let $X = \bigcup_1^k X_i$, $U = \bigcup_1^k U_i$ be disjoint decompositions, and let φ_i be isometries such that $\varphi_i(X_i) = U_i$ $(i = 1, \ldots, k)$. Similarly, let $Y = \bigcup_1^m Y_j$ and $V = \bigcup_1^m V_j$ be disjoint decompositions such that $\psi_j(Y_j) = V_j$ $(j = 1, \ldots, m)$, where $\psi_1, \ldots, \psi_m$ are isometries. We define $f(x) = \varphi_i(x)$ if $x \in X_i$ $(i = 1, \ldots, k)$, and $g(y) = \psi_j(y)$ if $y \in Y_j$ $(j = 1, \ldots, m)$. Then f is an injective map from X into Y, and g is an injective map from Y into X. By the Cantor–Bernstein–Schröder–Banach theorem, there are disjoint decompositions $X = A_1 \cup A_2$ and $Y = B_1 \cup B_2$ such that $f(A_1) = B_1$ and $g(B_2) = A_2$. Then

$$X = A_1 \cup A_2 = A_1 \cup g(B_2) = \bigcup_{i=1}^{k}(X_i \cap A_1) \cup \bigcup_{j=1}^{m}\psi_j(Y_j \cap B_2)$$

and

$$Y = B_1 \cup B_2 = f(A_1) \cup B_2 = \bigcup_{i=1}^{k}\varphi_i(X_i \cap A_1) \cup \bigcup_{j=1}^{m}(Y_j \cap B_2),$$

showing $X \equiv_{k+m} Y$.

The following theorem (the so-called "Banach–Tarski paradox") was discovered by S. Banach and A. Tarski in 1924.

Let B_1 and B_2 be disjoint three-dimensional balls of the same size. Then $B_1 \equiv_{10} (B_1 \cup B_2)$.

Proof. Let O and S denote the centre and the surface of B_1, respectively. If $x \in S$, then r_x will denote the segment with endpoints O and x, containing x but not containing O.

For every $C \subset S$ we denote $C^* = \bigcup_{x \in C} r_x$ (i.e., C^* is the "cone" with "base" C and vertex O minus the point O).

In the last section we proved that there is a set $A \subset S$ such that S contains infinitely many pairwise disjoint congruent copies of A, and S can be covered by four congruent copies of A. Let $S = \bigcup_{i=1}^{4} A_i$, where $A_i \simeq A$ $(i = 1, \ldots, 4)$. We put $C_1 = A_1$ and $C_i = A_i \setminus \bigcup_{j<i} A_j$ $(i = 2, 3, 4)$; then $S = \bigcup_{i=1}^{4} C_i$ and the sets C_i are pairwise disjoint. Therefore,

$$B_1 \setminus \{O\} = \bigcup_{i=1}^{4} C_i^*$$

is a disjoint decomposition. If γ denotes the translation mapping B_1 onto B_2, then

$$[C_1^* \cup \{O\}] \cup \left[\bigcup_{i=2}^{4} C_i^*\right] \cup \left[\bigcup_{i=1}^{4} \gamma(C_i^*)\right] \cup \{\gamma(O)\}$$

is a decomposition of $B_1 \cup B_2$ into $1 + 3 + 4 + 1 = 9$ disjoint subsets.

Let $D_i \subset S$ $(i = 1, 2, \ldots)$ be disjoint sets such that $D_i \simeq A$ for every i. Then $C_1^* \cup \{O\}$ is congruent to a subset of $D_1^* \cup \{O\}$; for $i = 2, 3, 4$, the set C_i^* is congruent to a subset of D_i^*; for $i = 1, \ldots, 4$, the set $\gamma(C_i^*)$ is congruent to a subset of D_{i+4}^*, and $\{\gamma(O)\}$ is congruent to any point of

D_9^*. Consequently, there is a subset $V \subset B_1$ such that $B_1 \cup B_2 \equiv_9 V$. Now $B_1 \equiv_1 B_1 \subset B_1 \cup B_2$ and thus, by the previous theorem, $B_1 \cup B_2 \equiv_{10} B_1$.

A similar argument proves the following theorem.

Let $B_1, \ldots, B_n$ be pairwise disjoint 3-dimensional balls of the same size. Then $B \equiv_{5n} B_1 \cup \cdots \cup B_n$.

Now we come to the main result of this section.

If $X, Y \subset \mathbf{R}^3$ are bounded and if there are balls (of arbitrary size) $B_1 \subset X$ and $B_2 \subset Y$, then $X \equiv Y$.

Proof. Since X is bounded, $X \subset \bigcup_{i=1}^n D_i$, where $D_i \simeq B_2$ $(i = 1, \ldots, n)$. Let $E_1, \ldots, E_n$ be disjoint balls congruent to B_2. Let $X_1 = X \cap D_1$ and $X_i = X \cap (D_i \setminus \bigcup_{j<i} D_j)$ $(i = 2, \ldots, n)$. Then $X = \bigcup_{i=1}^n X_i$ is a disjoint decomposition and $X_i \subset D_i$. Hence X is equidecomposable to a subset of $\bigcup_{i=1}^n E_i$. By the previous theorem, $\bigcup_{i=1}^n E_i \equiv B_2$ and hence there is a set $U \subset B_2$ such that $X \equiv U \subset Y$. Similarly, $Y \equiv V$ for a subset $V \subset B_1 \subset X$. Therefore we have $X \equiv Y$.

Note the following special cases of the previous theorem.

If B_1, B_2 are balls or cubes or polyhedra of arbitrary size, then $B_1 \equiv B_2$. For example, if C is a cube and D is a tetrahedron, then $C \equiv D$.

It follows that *there is no invariant and normed measure on all subsets of* $\mathbf{R}^3$. Indeed, suppose that m is such a measure. Then $m(X) = m(Y)$ holds whenever $X \equiv Y$. If Q is the unit cube $[0,1] \times [0,1] \times [0,1]$ and Q' is the union of two disjoint congruent copies of Q then, by the previous theorem, $Q \equiv Q'$. Hence $1 = m(Q) = m(Q') = 2$, a contradiction.

In the last section we remarked that the area of polygons can be extended to all Borel sets as an invariant and additive (even σ-additive) measure. This is true also in $\mathbf{R}^3$: the volume of polyhedra can be extended to all Borel sets as an invariant measure. Therefore, if A and B are polyhedra of different volumes, then the pieces used in $A \equiv B$ cannot all be Borel sets. But what if A and B are of the same volume? Are they equidecomposable using Borel pieces? This problem is unsolved even in the special case when A is a cube and B is a regular tetrahedron. This case is particularly interesting since, as was proven in Section 7, the cube and the regular tetrahedron are not equidecomposable in the geometric sense; that is, the pieces used in the decompositions cannot all be polyhedra.

Exercises

13.1. Prove that there is a bounded set $E \subset \mathbf{R}^2$ and there is a line segment I contained in E such that $E \equiv E \setminus I$. (H)

13.2. Suppose that the set $A \subset \mathbf{R}^2$ contains a disc. Prove that if I is any line segment, then $A \equiv A \cup I$. (H)

13.3. Prove that if $P \subset \mathbf{R}^2$ is a polygon and B is the boundary of P, then $P \equiv P \setminus B$.

13.4. Prove that if P_1 and P_2 are polygons of the same area, then $P_1 \equiv P_2$.

13.5. Prove that if $A \subset B \subset C$ and $A \equiv C$, then $A \equiv B$.

13.6. Prove that if $A \subset \mathbf{R}$ is bounded, then $\mathbf{R} \equiv \mathbf{R} \setminus A$.

13.7. Prove that if $A \subset \mathbf{R}^n$ is bounded, then $\mathbf{R}^n \equiv \mathbf{R}^n \setminus A$.

In the following exercises $X \equiv_\infty Y$ will denote the following statement: there are disjoint decompositions into countably many parts $X = \bigcup_{i=1}^\infty X_i$ and $Y = \bigcup_{i=1}^\infty Y_i$ such that $X_i \simeq Y_i$ for every i.

13.8. Prove that $\equiv_\infty$ is an equivalence-relation.

13.9. Prove that if $X \equiv_\infty U \subset Y$ and $Y \equiv_\infty V \subset X$, then $X \equiv_\infty Y$. (H)

13.10. Prove that $[0,1] \equiv_\infty \mathbf{R}$. (H)

13.11. Prove that if the sets $A, B \subset \mathbf{R}$ contain intervals (of arbitrary size), then $A \equiv_\infty B$.

13.12. Prove that if the sets $A, B \subset \mathbf{R}^n$ contain balls (of arbitrary size), then $A \equiv_\infty B$.

14

Open and Closed Sets in R. The Cantor Set

By an *interval* (in the wide sense) we shall mean a connected subset of $\mathbf{R}$. That is, a set $I \subset \mathbf{R}$ is an interval if $x < y < z$ and $x \in I$, $z \in I$ imply that $y \in I$. The empty set and the singletons are intervals (they are called *degenerate* intervals). The nondegenerate intervals are the following.

$$(a, b) = \{x \in \mathbf{R} : a < x < b\},$$

$$[a, b) = \{x \in \mathbf{R} : a \leq x < b\},$$

$$(a, b] = \{x \in \mathbf{R} : a < x \leq b\},$$

$$[a, b] = \{x \in \mathbf{R} : a \leq x \leq b\} \qquad \text{for every } a < b,$$

$$(-\infty, a) = \{x \in \mathbf{R} : x < a\},$$

$$(-\infty, a] = \{x \in \mathbf{R} : x \leq a\},$$

$$(a, \infty) = \{x \in \mathbf{R} : x > a\},$$

$$[a, \infty) = \{x \in \mathbf{R} : x \geq a\} \qquad \text{for every } a,$$

and finally, $\mathbf{R}$ is also an interval.

The maximal intervals contained in a set $H \subset \mathbf{R}$ are called the *components* of H. Every point $x \in H$ is contained in a component of H. Indeed, let I be the union of all intervals contained in H and containing x. It is easy to see that I is an interval, and I is clearly maximal. If I and J are distinct components of H, then I and J are disjoint. For, if $I \cap J \neq \emptyset$, then $I \cup J$ would also be an interval, contradicting the maximality of I and J. Therefore *every subset of $\mathbf{R}$ is the disjoint union of its components.* (Some of the components can be degenerate.)

A set $G \subset \mathbf{R}$ is called *open,* if every point $x \in G$ has a neighbourhood $(x - \delta, x + \delta)$ contained in G. The empty set, the intervals (a, b), $(-\infty, a)$, (a, ∞), and $\mathbf{R}$ itself are open; the other intervals are not. Deleting the endpoints of an interval I we obtain the *interior* of I. The interior of an interval is always open.

It is easy to see that the components of an open set must be open. Indeed, let I be one of the components of the open set G, and let $x \in I$. Then $x \in G$, and, since G is open, there is $\delta > 0$ such that $(x - \delta, x + \delta) \subset G$. By the maximality of I this implies $(x - \delta, x + \delta) \subset I$, which proves that I is open. Therefore every open set is the disjoint union of open intervals. Now every system S of disjoint open intervals is countable. Indeed, each nonempty open interval contains rational numbers. Selecting a rational number from each nonempty interval of the system S, we select distinct numbers from distinct intervals. Since the set of rationals is countable, this shows that the system S must be also countable. Putting these facts together, we obtain the following theorem.

Every open set in $\mathbf{R}$ is the disjoint union of countably many open intervals.

The structure of components of an open set can be rather complicated. Consider the following example. Let D_7 denote the set of those numbers $x \in (0, 1)$ whose decimal expansion contains the digit 7. If x has a terminating decimal expansion, then we require that both the terminating and the nonterminating expansions of x contain the digit 7. For example, the number 0.17 does not belong to D_7, since its nonterminating decimal expansion $0.16999\ldots$ contains no 7. On the other hand, $0.717 = 0.716999\ldots \in D_7$.

It is easy to see that D_7 is open. Indeed, let $x \in D_7$, and let the nonterminating decimal expansion of x be $0.a_1 a_2 \ldots$. If k is the first index with $a_k = 7$, then, as the expansion is non-terminating, there is an $n > k$ such that $a_n \neq 0$. Thus $(x - 10^{-n}, x] \subset D_7$. Since the number $0.a_1 \ldots a_{k-1}8$ does not contain the digit 7, there must be an index $m > k$ with $a_m < 9$, and thus $[x, x + 10^{-m}) \subset D_7$. Therefore $(x - \delta, x + \delta) \subset D_7$ for a suitable $\delta > 0$, and hence D_7 is open.

Let $b_1, \ldots, b_k$ be digits different from 7. If $a = 0.b_1 \ldots b_k 7$ and $b = 0.b_1 \ldots b_k 8$, then $a, b \notin D_7$ and $(a, b) \subset D_7$; that is, (a, b) is a component of D_7. One can prove (see Exercise 14.1) that every component of D_7 is of this form. This implies that *there are no adjacent components of D_7;* if (a_1, b_1) and (a_2, b_2) are components such that $b_1 < a_2$, then there is a component (a_3, b_3) such that $b_1 < a_3 < b_3 < a_2$ (see Exercise 14.2).

Now we turn to the closed sets. A set $H \subset \mathbf{R}$ is called *closed*, if $\mathbf{R} \setminus H$ is open. We show that *H is closed if and only if the limit of every convergent sequence of elements of H is also an element of H.* Indeed, let H be closed, and let $x_n \in H$, $x_n \to x$. If $x \notin, H$ then, as $\mathbf{R} \setminus H$ is open, we have $(x - \delta, x + \delta) \subset \mathbf{R} \setminus H$ with a suitable $\delta > 0$. In this case, however, $x_n \in (x - \delta, x + \delta) \subset \mathbf{R} \setminus H$ for every n large enough, contradicting $x_n \in H$. This proves the necessity of the condition. To prove sufficiency, suppose that $x_n \in H$, $x_n \to x$ implies $x \in H$. We prove that $\mathbf{R} \setminus H$ is open. Let $y \in \mathbf{R} \setminus H$; we have to show that

$$(y - \delta, y + \delta) \subset \mathbf{R} \setminus H$$

with a suitable $\delta > 0$. Suppose this is not true. Then

$$\left(y - \frac{1}{n}, y + \frac{1}{n} \right) \cap H \neq \emptyset$$

for every n. Let

$$x_n \in \left(y - \frac{1}{n}, y + \frac{1}{n} \right) \cap H \quad (n = 1, 2, \ldots).$$

Then $x_n \in H$ and $x_n \to y \notin H$, contradicting our assumption.

It is easy to see that the empty set and the singletons (more generally, the finite sets) are closed. The intervals $[a, b]$, $[a, \infty)$, $(-\infty, a]$, and $\mathbf{R}$ are also closed, and these are the only nondegenerate closed intervals. Note that $\emptyset$ and $\mathbf{R}$ are open and closed simultaneously. (These are the only subsets of $\mathbf{R}$ with this property; see Exercise 14.4.) The intervals $[a, b)$ and $(a, b]$ are neither open nor closed.

The famous *Cantor ternary set* (briefly Cantor set) is constructed as follows. Delete the open middle third of the interval $[0, 1]$. Then delete the open middle third of each of the remaining intervals $[0, 1/3]$ and $[2/3, 1]$. Then delete the open middle third of each of the remaining four intervals $[0, 1/9]$, $[2/9, 1/3]$, $[2/3, 7/9]$, and $[8/9, 1]$. Continue this process indefinitely. The set of the remaining points is called the Cantor ternary set, denoted by C. The set C is closed, since its complement is open, being the union of the half-lines $(-\infty, 0)$, $(1, \infty)$ and of those open intervals we have deleted during the process.

The elements of C can easily be recognized using their ternary expansions. (By the ternary expansion of a number we mean its expansion in base 3, when the digits are 0, 1, and 2.) In the first step of the construction we deleted from $[0, 1]$ the interval $(1/3, 2/3)$; that is, we deleted those numbers whose first digit in their ternary expansion is 1. In the second step we deleted from the remaining points the numbers whose second digit is 1, etc. We can see that a number $x \in [0, 1]$ belongs to C if and only if its ternary expansion contains the digits 0 and 2 only. (If x has a terminating expansion, then $x \in C$ if one of the two expansions contains only 0's and 2's. For example, $1/3 \in C$ because, in base 3, we have $1/3 = 0.0222\ldots$, and $2/3 \in C$ because $2/3 = 0.2$.)

Therefore, C is of the power of the continuum. Indeed, the map

$$(\varepsilon_1, \varepsilon_2, \ldots) \mapsto 0.(2\varepsilon_1)(2\varepsilon_2)\ldots$$

gives a bijection between the set of 0-1 sequences and the set C.

The set C is of measure zero; that is, it can be covered by a system of intervals with total length smaller than any prescribed positive number. Indeed, in the nth step of the construction we obtain 2^n closed intervals of length 3^{-n}. These intervals cover C, and their total length is $2^n \cdot (1/3^n) = (2/3)^n \to 0$ as $n \to \infty$.

An important and surprising property of C is that it can be mapped, using a continuous function, onto an interval. The proof follows. If $x \in C$ and if the ternary expansion of x is $0.a_1 a_2 \ldots$ ($a_i = 0$ or 2 for every i), then we define

$$f(x) = \sum_{i=1}^{\infty} \frac{a_i}{2^{i+1}}.$$

That is, $f(x)$ is obtained by dividing the digits of the ternary expansion of x by 2, and computing it as a binary expansion. It is clear that f maps C onto $[0, 1]$. It is also easy to see that f is monotone increasing on C. Therefore, if (a, b) is a component of the open set $[0, 1] \setminus C$, then $f(a) = f(b)$. Indeed, $f(a) < f(b)$ would imply that f could not take any value lying between $f(a)$ and $f(b)$, which is impossible. Now we extend f to $[0, 1]$ such that in every component (a, b) of $[0, 1] \setminus C$ we define f as the constant $f(a) = f(b)$. For the sake of simplicity, we also denote the extended function by f. This function is called the *Cantor function*. (Some authors call it "the Devil's staircase".)

The function is increasing, and maps $[0, 1]$ onto itself. This implies that f is continuous everywhere in $[0, 1]$. Indeed, let $x_0 \in (0, 1)$ and $\varepsilon > 0$ be given. Since f takes every value in $[0, 1]$, there are points $0 \le x_1 < x_0 < x_2 \le 1$ such that $f(x_1) = \max(f(x_0) - \varepsilon, 0)$ and $f(x_2) = \min(f(x_0) + \varepsilon, 1)$. Then, by monotonicity, $f(x_0) - \varepsilon < f(x) < f(x_0) + \varepsilon$ for every $x_1 < x < x_2$, proving the continuity of f at the point x_0. Similar arguments show that f

is continuous from the right at the point 0, and is continuous from the left at the point 1.

The name "Devil's staircase" refers to the property that $f(0) < f(1)$ in spite of the fact that $f'(x) = 0$ everywhere, apart from a set of measure zero. Indeed, if $x \in [0,1] \setminus C$, then f is constant in a neighborhood of x, which implies $f'(x) = 0$.

The ternary expansions $0.020202\ldots = 1/4$ and $0.202020\ldots = 3/4$ show that $1/4$ and $3/4$ are both elements of C. Together with 0 and 1, these give four dyadic rational numbers (that is, rationals of the form $p/2^k$) belonging to C. One can prove that there are no others (see Exercise 14.9).

Let $H \subset \mathbf{R}$ be a closed set. The components of the open set $\mathbf{R} \setminus H$ are called the *intervals contiguous to* H. If (a,b) and (b,c) are adjacent intervals contiguous to H, then b is an *isolated point of* H. It is clear that x is an isolated point of H if and only if there is a $\delta > 0$ such that $(x - \delta, x + \delta) \cap H = \{x\}$.

The Cantor set has no isolated points. This follows from the construction of its contiguous intervals, or can be seen from the ternary expansion of its elements. A closed set without isolated point is called *perfect*. Thus the Cantor set and the complement of the set D_7 are both perfect. The empty set and $\mathbf{R}$ are perfect as well.

We show next that *the cardinality of every nonempty perfect set is that of the continuum.*

Let $H \subset \mathbf{R}$ be a nonempty perfect set. We shall construct a closed interval $I_{\varepsilon_1 \ldots \varepsilon_n}$ for every finite 0-1 sequence $(\varepsilon_1, \ldots, \varepsilon_n)$ with the following properties:

(i) the interior of $I_{\varepsilon_1 \ldots \varepsilon_n}$ intersects H;

(ii) $I_{\varepsilon_1 \ldots \varepsilon_n}$ is shorter than $1/n$;

(iii) if $(\varepsilon_1 \ldots \varepsilon_n)$ and $(\eta_1 \ldots \eta_n)$ are distinct 0-1 sequences of the same length, then $I_{\varepsilon_1 \ldots \varepsilon_n} \cap I_{\eta_1 \ldots \eta_n} = \emptyset$ and

(iv) $I_{\varepsilon_1 \ldots \varepsilon_n} \supset I_{\varepsilon_1 \ldots \varepsilon_n \varepsilon_{n+1}}$ for every $\varepsilon_1, \ldots, \varepsilon_{n+1} \in \{0,1\}$.

Since H is nonempty and has no isolated points, it follows that H has at least two elements. Then there are disjoint closed intervals I_0 and I_1 of length < 1 such that their interiors intersect H. Suppose that $n \geq 1$ and we have constructed the intervals $I_{\varepsilon_1 \ldots \varepsilon_n}$ for every $\varepsilon_1, \ldots, \varepsilon_n \in \{0,1\}$ satisfying (i), (ii), and (iii). Let $(\varepsilon_1, \ldots, \varepsilon_n)$ be a 0-1 sequence of length n. Since H has no isolated points, (i) implies that the interior of $I_{\varepsilon_1 \ldots \varepsilon_n}$ contains at least two elements of H. Thus we can select disjoint closed intervals $I_{\varepsilon_1 \ldots \varepsilon_n 0} \subset I_{\varepsilon_1 \ldots \varepsilon_n}$ and $I_{\varepsilon_1 \ldots \varepsilon_n 1} \subset I_{\varepsilon_1 \ldots \varepsilon_n}$ of length $< 1/(n+1)$ such that their interior intersects H. In this way we have defined the intervals $I_{\varepsilon_1 \ldots \varepsilon_n}$ for every n and $\varepsilon_1, \ldots, \varepsilon_n \in \{0, 1\}$.

For every infinite 0-1 sequence $(\varepsilon_1, \varepsilon_2, \ldots)$, the intervals $I_{\varepsilon_1}, I_{\varepsilon_1 \varepsilon_2}, \ldots$ form a nested sequence of closed intervals. Therefore the intersection

$$\bigcap_{n=1}^{\infty} I_{\varepsilon_1 \ldots \varepsilon_n}$$

is nonempty. Let $x_{\varepsilon_1 \varepsilon_2 \ldots}$ be an element of this intersection. Then $x_{\varepsilon_1 \varepsilon_2 \ldots} \in H$. Indeed, if $x_n \in H \cap I_{\varepsilon_1 \ldots \varepsilon_n}$, then $|x_n - x_{\varepsilon_1 \varepsilon_2 \ldots}| < 1/n$ by (ii), and thus $x_n \to x_{\varepsilon_1 \varepsilon_2 \ldots}$. Since H is closed, this implies $x_{\varepsilon_1 \varepsilon_2 \ldots} \in H$.

If the 0-1 sequences $(\varepsilon_1, \varepsilon_2, \ldots)$ and $(\eta_1, \eta_2, \ldots)$ are distinct, then there is an n such that $\varepsilon_n \neq \eta_n$. Since $x_{\varepsilon_1 \varepsilon_2 \ldots} \in I_{\varepsilon_1 \ldots \varepsilon_n}$ and $x_{\eta_1 \eta_2 \ldots} \in I_{\eta_1 \ldots \eta_n}$, it follows from (iii) that $x_{\varepsilon_1 \varepsilon_2 \ldots} \neq x_{\eta_1 \eta_2 \ldots}$. This implies that the set $X = \{x_{\varepsilon_1 \varepsilon_2 \ldots} : \varepsilon_n \in \{0, 1\} \ (n = 1, 2, \ldots)\}$ is of the power of the continuum, and hence so is H.

One can prove that every closed set H can be represented as $H = P \cup M$, where P is perfect and M is countable (see Exercise 14.13). Therefore, *every closed set is either countable or is of the power of the continuum*. Indeed, if $P = \emptyset$, then $H = M$ is countable, otherwise $|H| = c$.

Exercises

14.1. Prove that if (a, b) is a component of D_7, then there are digits $b_1, \ldots, b_k$ different from 7 such that $a = 0.b_1 \ldots b_k 7$ and $b = 0.b_1 \ldots b_k 8$.

14.2. Prove that $[0, 1] \setminus D_7$ has no isolated points.

14.3. Let G be the union of the intervals

$$\left(\frac{p}{q} - \frac{1}{4q^2}, \frac{p}{q} + \frac{1}{4q^2} \right),$$

where p/q runs through all rational numbers. Prove that no component of G can be longer than 1. (H)

14.4. Prove that if $H \subset \mathbf{R}$ is open and closed simultaneously, then either $H = \emptyset$ or $H = \mathbf{R}$. (H)

14.5. Prove that every number $x \in [0, 2]$ can be written in the form $y + z$, where $y, z \in C$.

The aim of the next four exercises is to prove that the only dyadic rationals contained in the Cantor set are 0, 1/4, 3/4, and 1. The notation

$a \equiv b \pmod{c}$ means that the integers a and b give the same remainder when divided by c. In other words, $a \equiv b \pmod{c}$ if $c \mid b - a$.

14.6.   Prove that if $n \geq 4$, then $3^{2^{n-3}} \equiv 2^{n-1} + 1 \pmod{2^n}$.

14.7.   Prove that if $n \geq 3$, then the remainders of $\pm 3^k$ ($k = 0, 1, \ldots, 2^{n-2} - 1$) when divided by 2^n include every odd number between 1 and $2^n - 1$. (H)

14.8.   **a.** Prove that if $x \in C$, then $1 - x \in C$ and $\{3x\} \in C$ ($\{3x\}$ denotes the fractional part of $3x$).

   b. Prove that if $k/2^n \in C$ where k is odd, then each of the numbers $1/2^n$, $3/2^n$, ..., $(2^n - 1)/2^n$ belongs to C.

14.9.   Prove that if $k/2^n \in C$ where k is odd, then $n \leq 2$. (H)

14.10. Are there elements of C (apart from 0 and 1) with a finite (terminating) decimal expansion? (H)

14.11. A point x is said to be a *point of condensation of a set* $H \subset \mathbf{R}$ if $H \cap (x - \delta, x + \delta)$ is uncountable for every $\delta > 0$. (The point x need not be an element of H.) The set of all points of condensation of H will be denoted by H^*. Prove that $H \setminus H^*$ is countable for every $H \subset \mathbf{R}$. (H)

14.12. Prove that H^* is perfect for every $H \subset \mathbf{R}$. (H)

14.13. Prove that every closed set is the union of a perfect set and a countable set. (H)

15

The Peano Curve

By a *plane curve* we mean a map γ from an interval $[a, b]$ into $\mathbf{R}^2$. (We imagine $[a, b]$ as an interval of time. The map describes the position of a moving point at each instant $t \in [a, b]$.) Let $f(t)$ and $g(t)$ be the coordinates of the image of t under the map γ. Then we write $\gamma = (f, g)$. We say that the curve γ is continuous, if f and g are continuous on $[a, b]$.

It was discovered by G. Peano in 1890 that *there are continuous curves mapping an interval onto a square*. A curve with this property is called a *Peano curve*. In this section we shall give two constructions of Peano curves.

I. We construct a continuous map from $[0, 1]$ onto the unit square $Q = [0, 1] \times [0, 1]$. The straight lines $x = k/2^n$ and $y = k/2^n$ $(k = 1, \ldots, 2^n - 1)$ divide Q into 4^n congruent squares of size $2^{-n} \times 2^{-n}$. Our first aim is to enumerate these squares as $Q_0^n, \ldots, Q_{4^n-1}^n$ in such a way that
  (i) for every n and $0 < i \leq 4^n - 1$, the squares Q_{i-1}^n and Q_i^n are adjacent; that is, share a common side, and
  (ii) for every n and $0 \leq i \leq 4^n - 1$, $Q_i^n = Q_{4i}^{n+1} \cup Q_{4i+1}^{n+1} \cup Q_{4i+2}^{n+1} \cup Q_{4i+3}^{n+1}$.

We put $Q_0^0 = Q$. Suppose that $n \geq 0$ and the enumeration $Q_0^n, \ldots, Q_{4^n-1}^n$ is given and satisfies (i). We divide Q_0^n into four congruent nonoverlapping squares, and enumerate them as $Q_0^{n+1}, \ldots, Q_3^{n+1}$ satisfying the following conditions: for every $i = 1, 2, 3$, the squares Q_{i-1}^{n+1} and Q_i^{n+1} are adjacent, and one of the sides of Q_3^{n+1} lies on the common side of Q_0^n and Q_1^n. Then we divide Q_1^n into four congruent non-overlapping squares, and enumerate them as $Q_4^{n+1}, \ldots, Q_7^{n+1}$ in such a way that for every $i = 4, 5, 6, 7$, the squares Q_{i-1}^{n+1} and Q_i^{n+1} are adjacent, and one of the sides of Q_7^{n+1} lies

15	12	11	10
14	13	8	9
1	2	7	6
0	3	4	5

FIGURE 7

on the common side of Q_1^n and Q_2^n. (A simple inspection shows that this is always possible.) We continue in this way and obtain the enumeration Q_i^{n+1} ($i = 0, \ldots, 4^{n+1} - 1$). It is clear that these enumerations satisfy (i) and (ii). Figure 7 shows some possible enumerations for $n = 1$ and $n = 2$.

Let $0.a_1 a_2 \ldots$ be the expansion of the number $t \in [0, 1]$ in base 4. Then it follows from (ii) that the squares $Q_{a_1}^1$, $Q_{4a_1+a_2}^2$, $Q_{4^2 a_1 + 4 a_2 + a_3}^3, \cdots$ form a nested sequence of squares. Let $\gamma(t)$ denote the (unique) element of their intersection. Then γ maps $[0, 1]$ into Q.

If $x \in Q$ is arbitrary, then there is a nested sequence $Q_{i_1}^1 \supset Q_{i_2}^2 \supset \cdots$ such that $\bigcap_{n=1}^{\infty} Q_{i_n}^n = \{x\}$. By (ii), there is a sequence of digits $a_n = 0, 1, 2, 3$ such that $i_n = 4^{n-1} a_1 + 4^{n-2} a_2 + \cdots + a_n$ for every n. Then it follows from the definition of γ that $\gamma(0.a_1 a_2 \ldots) = x$. Since $x \in Q$ was arbitrary, this proves that γ maps $[0, 1]$ *onto* Q.

Now we prove that γ is continuous. We shall prove that if $|t_2 - t_1| < 1/4^n$, then

$$|\gamma(t_2) - \gamma(t_1)| \leq 2\sqrt{2}/2^n. \tag{1}$$

Let $t_1 = 0.a_1 a_2 \ldots$ and $t_2 = 0.b_1 b_2 \ldots$. If $i = 4^{n-1} a_1 + 4^{n-2} a_2 + \cdots + a_n$ and $j = 4^{n-1} b_1 + 4^{n-2} b_2 + \cdots + b_n$, then $|t_2 - t_1| < 1/4^n$ implies $|i - j| \leq 1$ and thus Q_i^n and Q_j^n are either identical or adjacent. Since $\gamma(t_1) \in Q_i^n$ and $\gamma(t_2) \in Q_j^n$, this implies (1), taking into consideration that the diameters of Q_i^n and Q_j^n are $\sqrt{2}/2^n$.

Let $\varepsilon > 0$ be given. If n is so large that $2\sqrt{2}/2^n < \varepsilon$, then it follows from (1) that $|\gamma(t_2) - \gamma(t_1)| < \varepsilon$ whenever $|t_2 - t_1| < 1/4^n$. If $\gamma = (f, g)$, then this implies $|f(t_2) - f(t_1)| < \varepsilon$ and $|g(t_2) - g(t_1)| < \varepsilon$ for every t_1, t_2 with $|t_2 - t_1| < 1/4^n$, which proves the continuity of f, g and γ.

II. First we show that there is a continuous map from C onto $C \times C$. Let the ternary expansion of $x \in C$ be $x = 0.a_1 a_2 \ldots$, where $a_i \in \{0, 2\}$ for every i. We define $\phi(x) = 0.a_1a_3a_5\ldots$ and $\psi(x) = 0.a_2a_4a_6\ldots$. Then ϕ and ψ map C into itself, and they are continuous in the sense that if $x, y \in C$ and $|x - y| < 1/3^{2n}$, then $|\phi(x) - \phi(y)| \le 1/3^n$ and $|\psi(x) - \psi(y)| \le 1/3^n$. It is easy to see that the map $x \mapsto (\phi(x), \psi(x))$ $(x \in C)$ maps C onto $C \times C$.

In the last section we constructed a continuous function $f : [0, 1] \to [0, 1]$ such that $f(C) = [0, 1]$. Let $g_1(x) = f(\phi(x))$ and $g_2(x) = f(\psi(x))$ for every $x \in C$. Then $x \mapsto (g_1(x), g_2(x))$ maps C onto $[0, 1] \times [0, 1]$.

Finally, we extend g_1 and g_2 to $[0, 1]$ such that in every interval contiguous to C we define g_1 and g_2 linearly. It is easy to show that these extensions are continuous (see Exercise 15.5). Thus the map $x \mapsto (g_1(x), g_2(x))$ $(x \in [0, 1])$ is a Peano curve.

An obvious modification of the second construction gives a continuous map from $[0, 1]$ onto the unit cube of $\mathbf{R}^3$, or, for that matter, of $\mathbf{R}^n$ for any given n. Moreover, we can map $[0, 1]$ continuously onto the "infinite dimensional" unit cube in the following sense.

There are continuous functions $g_i : [0, 1] \to [0, 1]$ $(i = 1, 2, \ldots)$ such that for every sequence $x_i \in [0, 1]$ there is a $t \in [0, 1]$ with $g_i(t) = x_i$ $(i = 1, 2, \ldots)$.

The proof is similar to the second construction. For every $x \in C$ we define $\phi_1(x), \phi_2(x), \ldots$ simultaneously, using distinct digits of x. The only difference is that, instead of decomposing the digits of x into two disjoint parts (even and odd indices), we have to decompose them into infinitely many infinite subsets.

We define g_i on C by $g_i(x) = f(\phi_i(x))$ $(x \in C)$, and then we extend g_i to $[0, 1]$ linearly in each interval contiguous to C.

Exercises

15.1. The definition of $\gamma(t)$ of the first construction was not complete; we should have checked that if t has two expansions (that is, if one of its expansion is terminating), then the two expansions yield the same point $\gamma(t)$. Prove it. (H)

15.2. Let γ be the Peano curve given by the first construction. Prove that

$$|\gamma(t_2) - \gamma(t_1)| \le 6|t_2 - t_1|^{1/2}$$

for every $t_1, t_2 \in [0, 1]$. (H)

15.3. Let γ be an arbitrary Peano curve mapping $[0, 1]$ onto $[0, 1] \times [0, 1]$, and suppose that there are positive constants C and α such that

$$|\gamma(t_2) - \gamma(t_1)| \le C|t_2 - t_1|^\alpha$$

for every $t_1, t_2 \in [0, 1]$. Prove that $\alpha \le 1/2$. (H)

15.4. Prove that if f and g are continuously differentiable functions on $[0, 1]$, then $\gamma = (f, g)$ cannot be a Peano curve. (H)

15.5. Let $g : C \to \mathbf{R}$ be continuous in the following sense: for every $x \in C$ and $\varepsilon > 0$ there is a $\delta > 0$ such that $|g(y) - g(x)| < \varepsilon$ whenever $y \in C$ and $|y - x| < \delta$. Extend g to $[0, 1]$ such that if (a, b) is an interval contiguous to C, then g is linear on $[a, b]$. Prove that the extended function is continuous on $[0, 1]$. (H)

15.6. Let $\gamma = (f, g)$ be a Peano curve mapping $[0, 1]$ onto $[0, 1] \times [0, 1]$. Let $g_1 = f$ and $g_{i+1} = g_i \circ g$ $(i = 1, 2, \ldots)$. Prove that

$$x \mapsto \big(g_1(x), \ldots, g_n(x)\big) \quad (x \in [0, 1])$$

maps $[0, 1]$ onto the n-dimensional unit cube. (H)

15.7. Let $g_1, g_2, \ldots$ be as in the previous exercise. Prove that for every sequence $x_i \in [0, 1]$ there is a $t \in [0, 1]$ with $g_i(t) = x_i$ $(i = 1, 2, \ldots)$. (H)

16

Borel Sets

A subset of the real line is called a *Borel set* if it can be obtained from open sets by applications of countable unions, countable intersections and taking complements. A more formal definition is the following. Let $\mathcal{B}$ denote the smallest class of subsets of $\mathbf{R}$ such that $\mathcal{B}$ contains the open sets, if $B \in \mathcal{B}$ then $\mathbf{R} \setminus B \in \mathcal{B}$, and if $B_1, B_2 \ldots \in \mathcal{B}$ then $\bigcup_{n=1}^{\infty} B_n \in \mathcal{B}$ and $\bigcap_{n=1}^{\infty} B_n \in \mathcal{B}$. (In other words, $\mathcal{B}$ denotes the intersection of all classes of sets satisfying these conditions.) The elements of the system $\mathcal{B}$ are called Borel sets, or simply Borel.

Every closed set, being the complement of an open set, is Borel. All countable unions of closed sets are Borel, too. They are called F_σ sets. Every closed set is F_σ. The half-open interval $[a, b)$ is neither open nor closed. It is, however, F_σ, since $[a, b) = \bigcup_{n=1}^{\infty} [a, b - (1/n)]$. The open intervals are also F_σ, as shown by

$$(a, b) = \bigcup_{n=1}^{\infty} \left[a + (1/n), b - (1/n) \right], \quad (a, \infty) = \bigcup_{n=1}^{\infty} [a + (1/n), \infty),$$

and

$$(-\infty, a) = \bigcup_{n=1}^{\infty} \left(-\infty, a - (1/n) \right].$$

Therefore *every open set is F_σ*. Indeed, every open set is the union of countably many open intervals. Replacing these open intervals by countable unions of closed sets, we find that the open set itself is a countable union of closed sets. (Here we used the fact that countable unions of countable sets are countable.)

A set is called G_δ, if it is the intersection of countably many open sets. Every open set is G_δ. Applying the identity

$$\mathbf{R} \setminus \bigcup_{n=1}^{\infty} A_n = \bigcap_{n=1}^{\infty} (\mathbf{R} \setminus A_n)$$

we find that every closed set is G_δ. That is, if a set is open or closed, then it is simultaneously F_σ and G_δ.

Since every point is closed, it follows that every countable set is F_σ. Thus the complements of countable sets are G_δ. The simplest example of a set that is *not* G_δ is the set of rationals (see Exercise 16.2). This implies that the set of irrationals is not F_σ.

Next come the unions of countably many G_δ sets; these are called $G_{\delta\sigma}$. By taking countable intersections of F_σ sets we obtain the $F_{\sigma\delta}$ sets. If a set is F_σ or G_δ, then it is simultaneously $G_{\delta\sigma}$ and $F_{\sigma\delta}$ (why?).

It is a remarkable fact that most of the sets that occur in analysis belong to (at least) one of the classes $G_{\delta\sigma}$ and $F_{\sigma\delta}$. Some examples: The set of points where an arbitrary function $f : [a, b] \to \mathbf{R}$ is continuous, is G_δ. The set of points where a continuous function is differentiable, is always $F_{\sigma\delta}$. If the terms of the infinite series $\sum_{n=1}^{\infty} f_n$ are continuous, then the set of points where the series is convergent is again $F_{\sigma\delta}$ (see Exercise 16.3).

To take a different kind of example, let D_7^∞ denote the set of those decimal expansions $0.a_1 a_2 \ldots$ in which $a_i = 7$ holds for infinitely many i. This set is $F_{\sigma\delta}$. Indeed, let $A_i = \{0.a_1 a_2 \ldots : a_i = 7\}$. Then A_i is the union of finitely many intervals, and thus A_i is F_σ. It is easy to check that $D_7^\infty = \bigcap_{n=1}^{\infty} \bigcup_{i=n}^{\infty} A_i$ showing that D_7^∞ is $F_{\sigma\delta}$.

Or try the following apparently more complicated set. Call the number $0.a_1 a_2 \ldots$ *normal with respect to the digit* 7, if $\lim_{n\to\infty} A(n)/n = 1/10$, where $A(n)$ is the number of indices $i \in \{1, \ldots, n\}$ such that $a_i = 7$. Then the set of those numbers that are normal with respect to 7 is still $F_{\sigma\delta}$ (see Exercise 16.4).

Actually, it is rather difficult to exhibit a set in a constructive way that is not $G_{\delta\sigma}$ or $F_{\sigma\delta}$. In this section we shall consider the following, more general problem. Let $\mathcal{Q}_0$ and $\mathcal{P}_0$ denote the families of open and closed sets, respectively. If $n > 0$ and the classes $\mathcal{P}_{n-1}$ and $\mathcal{Q}_{n-1}$ have been defined, then we put

$$\mathcal{P}_n = \left\{ \bigcap_{i=1}^{\infty} A_i : A_i \in \mathcal{Q}_{n-1} \ (i = 1, 2, \ldots) \right\},$$

and

$$\mathcal{Q}_n = \left\{ \bigcup_{i=1}^{\infty} A_i : A_i \in \mathcal{P}_{n-1} \ (i = 1, 2, \ldots) \right\}.$$

This defines $\mathcal{P}_n$ and $\mathcal{Q}_n$ for every $n = 0, 1, \ldots$. (Clearly, $\mathcal{P}_1, \mathcal{Q}_1, \mathcal{P}_2, \mathcal{Q}_2$ are the families of $G_\delta, F_\sigma, F_{\sigma\delta}$ and $G_{\delta\sigma}$ sets.) Now the problem is to construct a set that belongs to $\mathcal{P}_n$ but not to the classes of lower indices. This problem was solved by H. Lebesgue in 1905. We shall present the solution in this and the next section.

We shall need the notion of open and closed sets in $\mathbf{R}^2$. A set $H \subset \mathbf{R}^2$ is called open if, whenever $x \in H$, then H contains an open disc with centre x.

A set $H \subset \mathbf{R}^2$ is called closed if $\mathbf{R}^2 \setminus H$ is open. (The set $H \subset \mathbf{R}^2$ is closed if and only if $x_n \in H$ and $|x_n - x| \to 0$ imply $x \in H$.) The classes $\mathcal{P}_n$ and $\mathcal{Q}_n$ can be defined for subsets of $\mathbf{R}^2$ in the same way as for those of $\mathbf{R}$. In order to distinguish between the corresponding classes, we shall denote these classes of plane sets by $\mathcal{P}_n^2$ and $\mathcal{Q}_n^2$.

Let $H \subset \mathbf{R}^2$. The vertical section of H above the point x is defined by

$$H^x = \left\{ y \in \mathbf{R} : (x, y) \in H \right\} \quad (x \in \mathbf{R}).$$

It is easy to see that *if $H \in \mathcal{P}_n^2$ ($H \in \mathcal{Q}_n^2$), then $H^x \in \mathcal{P}_n$ ($H^x \in \mathcal{Q}_n$) for every $x \in \mathbf{R}$.*

Indeed, for the class $\mathcal{Q}_0^2$ this follows immediately from the definition of open sets in $\mathbf{R}$ and $\mathbf{R}^2$. By taking complements, we obtain the statement for $\mathcal{P}_0^2$. For arbitrary n an easy induction gives the statement.

A set $H \subset [0, 1] \times [0, 1]$ is called a *universal $\mathcal{P}_n$ set* if $H \in \mathcal{P}_n^2$, and for every $A \subset [0, 1]$ with $A \in \mathcal{P}_n$ there is an $x \in [0, 1]$ with $H^x = A$. The definition of universal $\mathcal{Q}_n$ sets is similar.

Our next aim is to prove the existence of universal sets. First we need the following simple lemma.

Let $\phi(x, y) = (f(x), g(y))$, where f and g are continuous functions defined on $[0, 1]$. If $H \subset [0, 1] \times [0, 1]$ and $H \in \mathcal{P}_n^2$ ($n \geq 0$), then

$$\phi^{-1}(H) = \left\{ (x, y) : \phi(x, y) \in H \right\} \in \mathcal{P}_n^2.$$

Proof. The inverse image $\phi^{-1}(H)$ of a closed set H is closed. (This is an immediate consequence of the continuity of f and g.) We proceed by induction, using the equalities

$$\phi^{-1}\left(\mathbf{R}^2 \setminus A \right) = \left([0, 1] \times [0, 1] \right) \setminus \phi^{-1}(A),$$

and

$$\phi^{-1}\left(\bigcup_{i=1}^{\infty} A_i\right) = \bigcup_{i=1}^{\infty} \phi^{-1}(A_i).$$

Now we turn to the construction of universal $\mathcal{P}_n$ sets and universal $\mathcal{Q}_n$ sets.

In the last section we constructed continuous functions $g_i : [0,1] \to [0,1]$ ($i = 1,2,\ldots$) such that for every sequence of numbers $x_i \in [0,1]$ there is an $x \in [0,1]$ with $g_i(x) = x_i$ ($i = 1,2,\ldots$). The construction also gave $g_i(0) = 0$ and $g_i(1) = 1$ for every i. We define

$$U = \big\{(x,y) \in (0,1) \times (0,1) :$$

$$\text{there is an } i \text{ such that } g_{2i-1}(x) < y < g_{2i}(x)\big\}.$$

We prove that U is a universal open set. First we show that U is open. Let $(x,y) \in U$; we show that there is an open rectangle containing (x,y) and lying in U. Since $(x,y) \in U$, we have $g_{2i-1}(x) < y < g_{2i}(x)$ for a suitable i. Let $\varepsilon > 0$ be chosen such that

$$g_{2i-1}(x) < y - \varepsilon < y + \varepsilon < g_{2i}(x).$$

By the continuity of the functions g_{2i-1} and g_{2i} there is a $\delta \in (0, \min(x, 1 - x))$ such that $g_{2i-1}(z) < y - (\varepsilon/2)$ and $y + (\varepsilon/2) < g_{2i}(z)$ for every $z \in (x - \delta, x + \delta)$. Then the open rectangle

$$(x - \delta, x + \delta) \times \big(y - (\varepsilon/2), y + (\varepsilon/2)\big)$$

lies in U. This proves that U is open.

Let $A \subset [0,1]$ be open. If $A = \emptyset$, then $U^0 = A$. If $A \neq \emptyset$ and $A = \bigcup_{i=1}^{\infty}(a_i, b_i)$, then let $x \in [0,1]$ be such that $g_{2i-1}(x) = a_i$ and $g_{2i}(x) = b_i$. It is easy to check that in this case $x \in (0,1)$ and $U^x = A$. This shows that U is a universal open set.

It is clear that $([0,1] \times [0,1]) \setminus U$ is a universal closed set. Now suppose that $n \geq 1$ and that V is a universal $\mathcal{P}_{n-1}$ set. Let $\phi_i(x,y) = (g_i(x), y)$. Then, by the last lemma, the set

$$V_i = \big\{(x,y) \in [0,1] \times [0,1] : y \in V^{g_i(x)}\big\}$$

$$= \big\{(x,y) \in [0,1] \times [0,1] : (g_i(x), y) \in V\big\}$$

belongs to $\mathcal{P}_{n-1}^2$, since $V_i = \phi_i^{-1}(V)$.

Thus $W = \bigcup_{i=1}^{\infty} V_i \in \mathcal{Q}_n^2$. We prove that W is a universal $\mathcal{Q}_n$ set. Let $A \subset [0,1]$ be a $\mathcal{Q}_n$ set. Then $A = \bigcup_{i=1}^{\infty} A_i$, where $A_i \in \mathcal{P}_{n-1}$ for every i. Since V is a universal $\mathcal{P}_{n-1}$ set, there are numbers $x_i \in [0,1]$ such that $V^{x_i} = A_i$ $(i = 1, 2, \ldots)$. Let $x \in [0,1]$ be such that $g_i(x) = x_i$ $(i = 1, 2, \ldots)$. Then

$$W^x = \bigcup_{i=1}^{\infty}(V_i)^x = \bigcup_{i=1}^{\infty}\{y : (x,y) \in V_i\}$$
$$= \bigcup_{i=1}^{\infty}V^{g_i(x)} = \bigcup_{i=1}^{\infty}V^{x_i} = \bigcup_{i=1}^{\infty}A_i = A.$$

Thus W is a universal $\mathcal{Q}_n$ set and then, obviously, $([0,1] \times [0,1]) \setminus W$ is a universal $\mathcal{P}_n$ set. This completes the proof.

Exercises

16.1. Prove that the set of Liouville numbers is G_δ.

16.2. Prove that the set of rational numbers is not G_δ. (H)

16.3. Let $f_n : [a,b] \to \mathbf{R}$ be continuous $(n = 1, 2, \ldots)$. Prove that the set of points $x \in [a,b]$ where $\sum_{n=1}^{\infty} f_n(x)$ converges is $F_{\sigma\delta}$.

16.4. Prove that the set of those numbers which are normal with respect to 7 is $F_{\sigma\delta}$.

16.5. Let $\mathcal{F}$ be an uncountable system of closed subsets of $\mathbf{R}$ such that for every $A, B \in \mathcal{F}$ we have either $A \subset B$ or $B \subset A$. Prove that $\bigcup \mathcal{F}$ is an F_σ set. (H)

16.6. Prove that $\mathcal{P}_n \cup \mathcal{Q}_n \subset \mathcal{P}_{n+1} \cap \mathcal{Q}_{n+1}$ for every $n = 0, 1, \ldots$.

16.7. Prove that if $A, B \in \mathcal{P}_n$, then $A \cup B$, $A \cap B \in \mathcal{P}_n$. Similarly, if $A, B \in \mathcal{Q}_n$ then $A \cup B$, $A \cap B \in \mathcal{Q}_n$.

16.8. Let $A_1, A_2, \ldots$ be Borel sets in $\mathbf{R}$. Prove that the set

$$\{x \in \mathbf{R} : \text{for every } n, \ x \in A_n \text{ implies } x \notin A_{n+1}\}$$

is also Borel.

16.9. Prove that every open set in $\mathbf{R}^2$ is the union of countably many closed squares. Infer that a set $H \subset \mathbf{R}^2$ is Borel if and only if H can be obtained from polygons by applications of countable unions, countable intersections, and taking complements. (This is how we defined the Borel subsets of $\mathbf{R}^2$ in Section 12).

17

The Diagonal Method

The so-called diagonal method, a technique used in certain constructions, was devised by Georg Cantor. The prototype is Cantor's second proof of the uncountability of the set of real numbers (see Section 10). In that argument we constructed, for any given sequence of real numbers $x_i = \pm n_i.a_1^i a_2^i \ldots$ $(i = 1, 2, \ldots)$ a number not contained in the sequence. The construction gives $x = 0.b_1 b_2 \ldots$, where $b_i \neq a_i^i$ for every i. This decimal expansion is formally different from each of the given decimal expansions of the numbers x_i. If $b_i \neq 0$ and $b_i \neq 9$ for every i, then the value of x will be different from each x_i. (In the actual construction we specified the values of b_i by defining $b_i = 5$ if $a_i^i \neq 5$ and $b_i = 4$ if $a_i^i = 5$.) The definition of x uses only the diagonal of the matrix $\left(a_j^i \right)$; this explains the name "diagonal method."

The essence of the method is even more transparent in the following argument. We prove that *the set A of all 0-1 sequences is uncountable.* (We already proved this in Section 10, where we showed that the set A is of the power of the continuum. The following direct proof is much simpler.)

We have to show that for every sequence $x_i \in A$ $(i = 1, 2, \ldots)$ there is an element of A which is different from each x_i. Let $x_i = (\varepsilon_1^i, \varepsilon_2^i, \ldots)$, where $\varepsilon_j^i \in \{0, 1\}$ for every i and j. Then $x = (1 - \varepsilon_1^1, 1 - \varepsilon_2^2, \ldots) \in A$ but $x \neq x_i$ $(i = 1, 2, \ldots)$.

A variant of the method works also for functions instead of sequences. Let us prove, for example, that *the set F of all functions $f : \mathbf{R} \to \mathbf{R}$ is not equivalent to $\mathbf{R}$.*

Suppose there exists a bijection mapping $\mathbf{R}$ onto F, and let ϕ_x denote the image of the number $x \in \mathbf{R}$ under this bijection. Let $f(x) = \phi_x(x) + 1$

for every $x \in \mathbf{R}$. Then $f \in F$, but f differs from each function ϕ_x ($x \in \mathbf{R}$), since $f(x) \neq \phi_x(x)$. This is impossible, since ϕ was a bijection from $\mathbf{R}$ onto F.

A similar argument works for the set P of all functions $f : \mathbf{R} \rightarrow \{0,1\}$. If $x \mapsto \phi_x$ is a map from $\mathbf{R}$ onto P, then we put $f(x) = 1 - \phi_x(x)$ for every $x \in \mathbf{R}$. Then $f \in P$ but $f \neq \phi_x$ for every x. That is, there is no bijection between $\mathbf{R}$ and P.

This proof actually shows that $\mathbf{R}$ *cannot be mapped onto F or P.* (We did not use the injectivity of the map in question.) Therefore *the cardinality of F (and of P) is greater than that of the continuum.*

The same argument works for any set X proving that *the cardinality of the set of all functions $f : X \rightarrow \{0,1\}$ is greater than that of X.* Since the functions $f : X \rightarrow \{0,1\}$ are exactly the characteristic functions of the subsets of X, we obtain that *the cardinality of the set of all subsets of X is greater than the cardinality of X.* If $P(X)$ denotes the set of all subsets of X, then the theorem states that $|P(X)| > |X|$ for every X. A direct proof, applying the diagonal method, runs as follows. Suppose that X can be mapped onto $P(X)$, and let $\Phi : X \rightarrow P(X)$ be such a map. Consider the set $A = \{x \in X : x \notin \Phi(x)\}$. Then $A \in P(X)$, and thus $A = \Phi(x)$ for some $x \in X$. If $x \in A = \Phi(x)$, then, by the definition of A, we have $x \notin A$. On the other hand, if $x \notin A = \Phi(x)$, then, again by the definition of A, we have $x \in A$. We obtain a contradiction in both cases, proving that X cannot be mapped onto $P(X)$.

Let Ω denote the set of all sets. Then every subset of Ω is, at the same time, an element of Ω and thus $P(\Omega) \subset \Omega$. This implies $|P(\Omega)| \leq |\Omega|$, contradicting the theorem just proved. Georg Cantor, the creator of set theory, was already aware of this contradiction in 1895. He commented on this antinomy by claiming that Ω is an "absolutely infinite and inconsistent collection". But Bertrand Russell observed in 1903 that applying the proof of $|P(X)| > |X|$ to $X = \Omega$ we obtain another, even more spectacular contradiction as follows. The identity maps Ω onto $P(\Omega)$ (in fact, it maps onto Ω which is strictly larger than $P(\Omega)$). Substituting the map $\Phi(x)$ by x in the proof, we arrive at the following contradiction. *Let us consider the set $A = \{x : x \notin x\}$. If $A \in A$, then, by the definition of A, we have $A \notin A$. If, on the other hand, $A \notin A$, then, again by the definition of A, we have $A \in A$.*

This is the so-called *Russell paradox.* Until the development of axiomatic set theory, this was considered a serious antinomy shattering the fundamentals of set theory and even those of mathematics as a whole. After the creation of axiomatic set theory, such paradoxes became harmless. In this theory we cannot form sets arbitrarily, only if it is allowed by the axioms. The axioms themselves postulate the existence of certain sets, or state the existence of some sets assuming the existence of others. Since the existence of Ω would imply a contradiction, we simply conclude that *no set can contain every set*. Similarly, Russell's argument implies that no set consists of those sets which are not elements of themselves.

Now we turn to the existence of Borel sets of arbitrary class. We prove, applying the diagonal method, that *for every n there exists a Borel set belonging to $\mathcal{P}_n$ but not to $\mathcal{Q}_n$*. Since $\mathcal{P}_{n-1} \cup \mathcal{Q}_{n-1} \subset \mathcal{Q}_n$ (see Exercise 16.6), this will prove the existence of a Borel set belonging to $\mathcal{P}_n \setminus (\mathcal{P}_{n-1} \cup \mathcal{Q}_{n-1})$.

For an arbitrary plane set $H \subset \mathbf{R}^2$ we shall denote

$$H^* = \{x \in \mathbf{R} : (x, x) \in H\}.$$

First we show that *if $H \in \mathcal{P}_n^2$ ($H \in \mathcal{Q}_n^2$), then $H^* \in \mathcal{P}_n$ ($H^* \in \mathcal{Q}_n$)*.

Indeed, for the class $\mathcal{Q}_0^2$ this easily follows from the definition of open sets in $\mathbf{R}$ and $\mathbf{R}^2$. By taking complements, we obtain the statement for $\mathcal{P}_0^2$. For arbitrary n an easy induction gives the statement, using

$$\left(\bigcup_{i=1}^{\infty} H_i \right)^* = \bigcup_{i=1}^{\infty} H_i^* \quad \text{and} \quad \left(\bigcap_{i=1}^{\infty} H_i \right)^* = \bigcap_{i=1}^{\infty} H_i^*.$$

Now let $H \subset [0,1] \times [0,1]$ be a universal $\mathcal{Q}_n$ set, and put

$$K = \{x \in [0,1] : (x, x) \notin H\} = [0,1] \setminus H^*.$$

Then $K \in \mathcal{P}_n$, since $H^* \in \mathcal{Q}_n$ by the previous argument. We prove that $K \notin \mathcal{Q}_n$. Indeed, suppose $K \in \mathcal{Q}_n$. Then, as H is a universal $\mathcal{Q}_n$ set, there is an $x \in [0,1]$ such that $K = H^x$. Then we have either $(x, x) \in H$ or $(x, x) \notin H$. In the first case we have $x \notin K$ by the definition of K on the one hand, but $x \in H^x = K$ on the other hand. In the second case we have $x \in K$ and $x \notin H^x = K$. Since both cases are impossible, this shows $K \notin \mathcal{Q}_n$, completing the proof.

Exercises

17.1. a. Prove that there exists a function of two variables $f : \mathbf{R}^2 \to \mathbf{R}$ such that, for every polynomial $p(x)$ of one variable, there is a $y \in \mathbf{R}$ such that $p(x) = f(x, y)$ for every $x \in \mathbf{R}$. (H)

b. Can f be a polynomial of two variables? (H)

17.2. Solve the previous exercise by replacing the word "polynomial" with "continuous". (H)

17.3. Prove that there exists a continuous function of two variables $f : \mathbf{R}^2 \to \mathbf{R}$ such that for every polynomial $p(x)$ of one variable there is a $y \in \mathbf{R}$ such that $p(x) = f(x, y)$ for every $x \in \mathbf{R}$. (H)

17.4. Let n be a positive integer. Prove that there are continuous functions $f_i : \mathbf{R} \to \mathbf{R}$ $(i = 1, \ldots, n)$ such that for every $x_1, \ldots, x_n \in \mathbf{R}$ there is an $x \in \mathbf{R}$ with $f_i(x) = x_i$ $(i = 1, \ldots, n)$.

17.5. Are there continuous functions $f_i : \mathbf{R} \to \mathbf{R}$ $(i = 1, 2, \ldots)$ such that, for every sequence x_i of real numbers, there is an $x \in \mathbf{R}$ with $f_i(x) = x_i$ $(i = 1, 2, \ldots)$? (H)

17.6. Does there exist a Borel set $H \subset [0, 1] \times [0, 1]$ such that for every Borel set $A \subset [0, 1]$ there is an $x \in [0, 1]$ with $H^x = A$?

17.7. Prove that there exists a Borel set $H \subset \mathbf{R}$ such that $H \notin \mathcal{P}_n$ for every $n = 1, 2, \ldots$. (H)

References

The monographs [6] and [12] provide good introductions into number theory. The book [11] discusses questions of irrationality and transcendence. As for algebraic numbers and fields, see [2], [17] and [18] (the latter is rather advanced). Cauchy's functional equation is studied in [1] and [8]. The standard monograph on questions of equidecomposability is [19]. The survey article [10] contains additional information on this topic. For more details concerning Sidon sequences see [4] and [5]. The book [9] gives the basics of set theory and topology.

Of the many books offering a general introduction to mathematics we mention [3], [7], [13], [14], [15] and [16]. The interested reader may find further references connected to some of the topics of this book in [13] and [15].

1. J. Aczél and J. Dhombres: *Functional equations in several variables. With applications to mathematics, information theory and to the natural and social sciences.* Encyclopedia of Mathematics and its Applications, 31. Cambridge University Press, 1989.

2. H. Cohn: *Advanced Number Theory.* Dover, 1980.

3. R. Courant and H. Robbins: *What is Mathematics?* Oxford University Press, 1973

4. R. K. Guy: *Unsolved Problems in Number Theory.* Second edition, Springer, 1994.

5. H. Halberstam and K. F. Roth: *Sequences.* Springer, 1983.

6. G. H. Hardy and E. M. Wright: *An Introduction to the Theory of Numbers.* Fifth edition, Clarendon Press (Oxford), 1979.

7. M. Kac and S. M. Ulam: *Mathematics and Logic*. The New American Library, 1969.

8. M. Kuczma: *An Introduction to the Theory of Functional Equations and Inequalities. Cauchy's Equation and Jensen's Inequality*. Państwowe Wydawnictwo Naukowe, Uniw. Ślaski, Warszawa–Kraków–Katowice, 1985.

9. K. Kuratowski: *Introduction to Set Theory and Topology*. Pure and Applied Mathematics Vol. 101. Second English edition, PWN and Pergamon Press, 1977.

10. M. Laczkovich, Paradoxical decompositions: a survey of recent results, in: *First European Congress of Mathematics (Paris, July 6–10, 1992)*. Progress in Mathematics, No. 120, Birkhäuser, 1994, Volume II, 159–184.

11. I. Niven: *Irrational Numbers*. The Carus Mathematical Monographs No. 11. The Mathematical Association of America, 1967.

12. I. Niven, H. S. Zuckerman and H. L. Montgomery: *An Introduction to the Theory of Numbers*. Fifth edition, John Wiley & Sons, 1991.

13. D. E. Penney: *Perspectives in Mathematics*. W. A. Benjamin, 1972.

14. H. Rademacher und O. Toeplitz: *Von Zahlen und Figuren*. Second edition, Springer, 1933. English translation: *The Enjoyment of Mathematics*. Dover, 1990.

15. S. K. Stein: *Mathematics, the Man-Made Universe*. Third edition, W. H. Freeman and Co., 1976.

16. H. Steinhaus: *Mathematical Snapshots*. Third American edition, Oxford University Press, 1983.

17. I. N. Stewart and D. O. Tall: *Algebraic Number Theory*. Chapman and Hall, 1979.

18. B. L. van der Waerden: *Algebra*. Springer, 1991.

19. S. Wagon: *The Banach–Tarski paradox*. First paperback edition, Cambridge University Press, 1993.

Hints

Section 1

1.5. Express $\sin 2r$ and $\cos 2r$ as rational functions of $\tan r$.

1.6. Use the expansion
$$\sin 1 = 1 - \frac{1}{3!} + \frac{1}{5!} - \cdots .$$
and apply the argument of the first proof of the irrationality of e.

1.7. Consider the integral
$$\int_0^1 \frac{1}{n!} x^n (1 - x)^n e^{2x} \, dx$$
and apply the argument used in the proof of the irrationality of π.

Section 2

2.7. Let r denote the radius of the circumscribed circle. Then $r \cos \alpha = 1$, $r \cos \beta = 2$, $r \cos \gamma = 3$. Applying the result of the previous exercise, show that r is the root of a cubic polynomial with integer coefficients and without rational roots.

Section 3

3.5. Determine the remainder of $2^{2^i} + 3 \bmod 7$.

Section 4

4.2.   Prove, by induction, the following stronger statement:

$$\sqrt{p_n} \notin \mathbf{Q}\left(\sqrt{p_1}\right) \cdots \left(\sqrt{p_{n-1}}\right).$$

4.3.   The condition is that $\sqrt{a_i/a_j}$ is irrational for every $i \neq j$.

Section 5

5.1.   **a.** Prove that $t_n - (-1)^k \cdot 2$ has a local extremum at the point $\cos(k\pi/n)$.

5.5.   Prove that $\mathbf{Q}(\tan 1^\circ) = \mathbf{Q}(\cos 2^\circ)$ by expressing $\cos 2\alpha$ as a rational function of $\tan \alpha$, and $\tan \alpha$ as a rational function of $\sin 2\alpha$ and $\cos 2\alpha$.

Section 6

6.5.   Prove that the limit

$$\lim_{n \to \infty} f(nx)/n = g(x)$$

exists for every x.

6.11.   Let

$$[a, b] \times [c, d] = R_1 \cup \cdots \cup R_n,$$

where $R_1, \ldots, R_n$ are nonoverlapping rectangles. Prove that there are points

$$a = a_0 < \cdots < a_k = b, \quad c = c_0 < \cdots < c_m = d$$

such that each R_i is of the form

$$[a_p, a_q] \times [c_r, c_s] \quad (0 \leq p < q \leq k, \ 0 \leq r < s \leq m).$$

Section 7

7.4.   **b.** The points $(\pm 1, 0, 0)$, $(0, \pm 1, 0)$, $(0, 0, \pm 1)$ are the vertices of an octahedron. The points $(1, 0, 0)$, $(0, 1, 0)$, $(0, 0, 1)$, $(1, 1, 1)$ are the vertices of a regular tetrahedron.

Section 8

8.3.   Let $a_i = b_i \cdot c_i$ where b_i is a power of two and c_i is odd. What are the possible values of b_i?

8.6.   Prove that $r_n^2 - 3s_n^2 = 1$.

Section 9

9.4.   If $\alpha = a/b \neq 0$, then try

$$\frac{p_n}{q_n} = \frac{na}{nb + 1}.$$

9.7.   Selecting a subsequence, we may assume that

$$(2c)q_n \leq q_{n+1} \leq (4c^2)q_n^2.$$

Let $q_{n-1} \leq q \leq q_n$. Prove that

$$\left| \alpha - \frac{p}{q} \right| > \frac{d^8}{q}$$

with a constant d using the estimate

$$\left| \alpha - \frac{p}{q} \right| \geq \left| \frac{p}{q} - \frac{p_{n+1}}{q_{n+1}} \right| - \left| \alpha - \frac{p_{n+1}}{q_{n+1}} \right|.$$

Section 10

10.5.   Every interval contains a rational number.

10.6. Every disc contains a point with rational coordinates.

10.15. $(1,1) = (-1,0] \cup (0,1)$.

10.17. The power set of $\mathbf{N}$ is equivalent to the set of all 0-1 sequences.

10.18. Each $x \in (0,1]$ has a unique infinite expansion $x = 2^{-a_1} + 2^{-a_2} + \cdots$, where $a_1 < a_2 < \cdots$ are natural numbers. Now use the bijections

$$x \leftrightarrow (a_1, a_2, \ldots) \leftrightarrow (a_1, a_2 - a_1, a_3 - a_2, \ldots).$$

10.19. We may assume that $A = B$ is the set of all 0-1 sequences. Now use the bijection

$$((n_1, n_2, \ldots), (k_1, k_2, \ldots)) \leftrightarrow (n_1, k_1, n_2, k_2, \ldots).$$

10.21. $\mathbf{R}^2 = \bigcup_{y \in \mathbf{R}} \{(x,y); \ x \in \mathbf{R}\}$ and $\mathbf{R}^2 \sim \mathbf{R}$.

10.28. Let f be a similarity transformation mapping A into B, let g be a similarity transformation mapping B into A, and apply the Cantor–Bernstein–Schröder–Banach theorem.

Section 11

11.4. Represent α and β as compositions of two reflections. By suitable "simultaneous manipulation" of the reflections, show that $\alpha\beta$ is also the composition of two reflections.

11.6. Let α and β be rotations about the intersecting lines ℓ_1 and ℓ_2, respectively. Let π be a plane containing both ℓ_1 and ℓ_2. Then there are planes π_1 and π_2 such that $\alpha = r_{\pi_1} r_\pi$ and $\beta = r_\pi r_{\pi_2}$. Therefore $\alpha\beta = r_{\pi_1} r_\pi r_\pi r_{\pi_2} = r_{\pi_1} r_{\pi_2}$, and thus $\alpha\beta$ is a rotation about the intersection of π_1 and π_2 (or is the identity if $\pi_1 = \pi_2$).

11.7. Let Λ denote the lattice points of the lattice obtained by tiling the plane with regular triangles of side length 1. Show that the points $f(x)$ $(x \in \Lambda)$ form a similar lattice.

11.8. Let ABC be a triangle with $\overline{AB} = 1$, $\overline{BC} = \overline{CA} = 2$. Prove that the image of ABC_Δ is a triangle congruent to ABC_Δ. Show that the midpoint of the segment AB is mapped onto the midpoint of the segment $f(A)f(B)$.

11.10. If $|x - y| < 1/2^k$, then there is a point z such that $|x - z| = |y - z| = 1/2^k$. This implies that if $|x - y| < 1/2^k$, then $|f(x) - f(y)| \leq 2/2^k$.

Section 12

12.1. Take, for example, $m(H) = 1$ if $0 \in H$ and $m(H) = 0$ if $0 \notin H$.

12.2. Let A be as in the proof of Vitali's theorem. Prove $m(A) = 0$ (without using monotonicity) by noting that the measure of every subset of $[0, 1]$ must be finite. From this infer that $m(\mathbf{R}) = 0$. This implies that the measure of every set is finite. Consider

$$H = \bigcup_{n=1}^{\infty} [2n - 1, 2n].$$

12.4. Prove that S contains infinitely many disjoint congruent copies of C.

12.5. Let $C \subset [a, b]$. Prove that $[a, b + 1]$ contains n disjoint congruent copies of C for every $n = 1, 2, \ldots$.

Section 13

13.1. Let α/π be irrational, and let I_n denote the segment obtained by rotating $[0, 1]$ about the origin by the angle $n\alpha$. Prove that

$$E = \bigcup_{n=0}^{\infty} I_n$$

satisfies the requirements.

13.2. Let $D \subset A$ be a disc. Suppose first that the radius of D is greater than the length of I and apply the previous exercise. If the radius of D is smaller than the length of I, then cut I into finitely many short segments.

13.9. Apply the Cantor–Bernstein–Schröder–Banach theorem.

13.10. Combine Vitali's construction with the previous exercise.

Section 14

14.3. Prove that $\sqrt{2} + n \notin G$ for every integer n.

14.4. Suppose that $H \neq \emptyset$ and $H \neq \mathbf{R}$. Let $a \in H$, $b \notin H$, assume $a < b$, and put $c = \sup H \cap [a, b]$. Prove that if $c \in H$, then H is not open, and if $c \notin H$, then H is not closed.

14.7. Let d denote the smallest positive integer such that $3^d \equiv 1 \pmod{2^n}$. Prove, using the previous exercise, that $d = 2^{n-2}$. Infer that the remainders of 3^k ($k = 0, 1, \ldots, 2^{n-2} - 1$) mod 2^n are different. Finally, observe that $3^k \not\equiv -1 \pmod 8$, and thus $3^i \not\equiv -3^j \pmod{2^n}$ for every i, j.

14.9. If $n \geq 3$, then one of the numbers $1/2^n, 3/2^n, \ldots, (2^n - 1)/2^n$ lies in $(1/3, 2/3) \subset \mathbf{R} \setminus C$.

14.10. Each of the numbers $1/10, 3/10, 7/10, 9/10$ belongs to C. Find others!

14.11. If $x \in H \setminus H^*$, then there are rational numbers $p < q$ such that $p < x < q$ and $[p, q] \cap H$ is countable. The union of these countable sets $[p, q] \cap H$ covers $H \setminus H^*$.

14.12. It is easy to see that H^* is closed, therefore it is enough to show that H^* has no isolated points. Suppose x is an isolated point, and let $\delta > 0$ be such that $H^* \cap (x - \delta, x + \delta) = \{x\}$. Then $H \cap (x - \delta, x)$ has no points of condensation and thus, by the previous exercise, it is countable. By the same reason, $H \cap (x - \delta, x)$ is countable, too, and then so is $H \cap (x - \delta, x + \delta)$. This, however, contradicts $x \in H^*$.

14.13. If H is closed, then $H^* \subset H$, and thus $H = H^* \cup (H \setminus H^*)$.

Section 15

15.1. Use the argument of the proof of (1).

15.2. If $t_1 \neq t_2$, then choose a positive integer n such that

$$\frac{1}{4^{n+1}} \leq |t_2 - t_1| < \frac{1}{4^n},$$

and apply (1).

15.3. Let

$$H_i = \gamma \left(\left[\frac{i-1}{n}, \frac{i}{n} \right] \right) \qquad (i = 1, \ldots, n).$$

Prove that H_i can be covered by a disc of radius C/n^α. Since

$$\bigcup_{i=1}^{n} H_i = [0,1] \times [0,1],$$

this implies $1 \le n \cdot (C/n^\alpha)^2 \pi$.

15.4. Show that if f and g are both continuously differentiable, then there is a constant $C > 0$ such that

$$\big|\gamma(t_2) - \gamma(t_1)\big| \le C \cdot |t_2 - t_1|$$

for every $t_1, t_2 \in [0,1]$.

15.5. It is clear that g is continuous at each point of $[0,1] \setminus C$. To prove that g is continuous from the right at a point $x \in C \cap [0,1)$, distinguish between the cases whether or not x is the left endpoint of an interval contiguous to C. If x is not such a point, and $\big|g(y) - g(x)\big| < \varepsilon$ whenever $y \in C$ and $|y - x| < \delta$, then put $x_1 \in C \cap (x, x+\delta)$, and prove that $\big|g(y) - g(x)\big| < \varepsilon$ for every $y \in (x, x_1)$.

15.6. Apply induction, starting with $n = 1$. Warning: the induction step is tricky!

15.7. Let $t_n \in [0,1]$ be such that $g_i(t_n) = x_i$ $(i = 1, \ldots, n)$, and let t be a point of accumulation of the sequence t_n.

Section 16

16.2. Suppose $\mathbf{Q} = \bigcap_n G_n$, where G_n is open. Find a nested sequence of closed intervals I_n such that $I_n \subset G_n$ and $\bigcap_n I_n$ consists of a single irrational number.

16.5. Let $\bigcup \mathcal{F} = F$. If F is closed, there is nothing to prove. If F is not closed, let $x_n \in F$, $x_n \to x$, $x \notin F$. Prove that if $x_n \in A_n \in \mathcal{F}$, then $F = \bigcup_{n=1}^{\infty} A_n$.

Section 17

17.1. **a.** Prove that the set of polynomials is of the power of the continuum.

b. If f is a polynomial, then so is $f(x,x) + 1$.

17.2. The set of all continuous functions defined on $\mathbf{R}$ is also of the power of the continuum. To prove this, observe that such a function is determined by its restriction to $\mathbf{Q}$.

17.3. Using Peano curves construct, for every n, a continuous function

$$f : \left(\mathbf{R} \times [2n - 1, 2n]\right) \to \mathbf{R}$$

such that if p is a polynomial of degree n with coefficients belonging to $[-n, n]$, then there is a $y \in [2n - 1, 2n]$ such that $p(x) = f(x, y)$ for every x.

17.5. Let $M_i = \max_{[-i,i]} \left| f_i(x) \right|$, and put $x_i = M_i + 1$.

17.7. If $H_n \subset (n, n + 1)$ and $H_n \notin \mathcal{P}_n$, then $\bigcup_n H_n \notin \bigcup_n \mathcal{P}_n$.

Index